John Ambrose Fleming

Electric Lamps and electric Lighting

A Course of four Lectures on electric Illumination delivered at the Royal institution

of Great Britain

John Ambrose Fleming

Electric Lamps and electric Lighting
A Course of four Lectures on electric Illumination delivered at the Royal institution of Great Britain

ISBN/EAN: 9783337069247

Printed in Europe, USA, Canada, Australia, Japan

Cover: Foto ©ninafisch / pixelio.de

More available books at **www.hansebooks.com**

SECOND EDITION.

ELECTRIC LAMPS

— AND —

ELECTRIC LIGHTING.

A Course of Four Lectures on Electric Illumination

DELIVERED AT THE

Royal Institution of Great Britain.

BY

J. A. FLEMING, M.A., D.Sc., F.R.S.

PROFESSOR OF ELECTRICAL ENGINEERING IN UNIVERSITY COLLEGE,
LONDON,

MEMBER OF THE INSTITUTION OF ELECTRICAL ENGINEERS.

*The Illustrations, Designs, and Text of these Lectures are entered at Stationers
Hall, and are, therefore, copyright.*

London:
"THE ELECTRICIAN" PRINTING & PUBLISHING COMPANY,
LIMITED,
SALISBURY COURT, FLEET STREET.
1899.

Works by Dr. J. A. Fleming.

Alternate-Current Transformer, Vol. I. *New Edition.*

Alternate-Current Transformer, Vol. II.

Magnets and Electric Currents.
  An Elementary Treatise for Electrical Artisans and Science Teachers.

Electrical Laboratory Notes and Forms.
  Consisting of 20 Elementary and 20 Advanced Papers for the use of Students in Electrical Engineering Classes. Intended as a help to the Teacher and his Assistants. A guide to the Student.

Centenary of the Electric Current, 1799-1899.

PREFACE
TO THE
SECOND EDITION.

THE following pages are a transcript, with considerable additions, of a course of Four Lectures on "Electric Illumination," delivered by the Author in 1894 to afternoon audiences at the Royal Institution, London. They are in no sense presented as a complete treatment of a subject which has long since outgrown the limits of a single small book or course of lectures. The original aim was merely to offer to a general audience such non-technical explanations of the physical effects and problems concerned in the modern applications of electricity for illuminating purposes as might serve to further an intelligent interest in the subject, and perhaps pave the way for a more serious study of it. In preparing a new issue of the book the Author has added to each chapter paragraphs which serve to bring up the information more into line with recent practice, and, without departing from the elementary character of the work, made the necessary corrections to preserve the text from being antiquated as to statement.

Many of those who were present at the Lectures expressed a desire to be again able to refer to the descriptions and illustrations then given, and at their request, and in the hope that they may be of use to others, the lecture notes have been revised and published. The aim throughout has been rather to deal with principles than with details, and to give such general and guiding explanations of terms and processes as are essential for a grasp of the outlines of the subject. For this purpose simple drawings have been given of as many of the experiments and apparatus used as possible. For permission to reproduce a set of photographs of views of the interior of rooms artistically illuminated with incandescent lamps, and which are to be found in the second Lecture, the author is indebted to Mr. A. F. Davies, under whose guidance these particular instances of electric lighting were carried out.

J. A. F.

University College, London.
November, 1899.

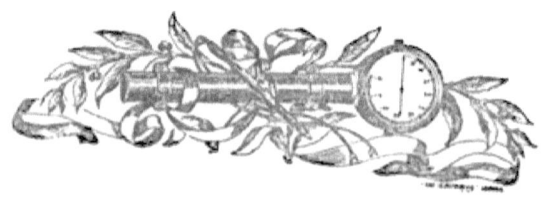

CONTENTS.

(For Index to Contents, see page 267.)

LECTURE I.
ELECTRIC MEASUREMENTS ... ... PAGES 1—50
14 Illustrations.

LECTURE II.
ELECTRIC GLOW LAMPS ... ... 51—137
40 Illustrations.

LECTURE III.
ELECTRIC ARC LAMPS ... ... 139—191
14 Illustrations.

LECTURE IV.
ELECTRIC DISTRIBUTION ... ... 193—263
39 Illustrations.

ERRATA.

The reader is requested to make the following corrections:—

Page 118, last line—
> *for* " Thompson "
> *read* " Thomson."

Page 133, seventh line from bottom—
> *for* " slight resistance "
> *read* " slight increase in the resistance."

Page 135, in the table—
> *for* " 26 and 25 "
> *read* " 2·6 and 2·5."

Page 142, seventh line from bottom—
> *for* " Thompson "
> *read* " Thomson."

LECTURE I.

ELECTRIC LIGHTING in Great Britain.—A Glance Backward over Twenty Years.—Present Condition of Electric Lighting—An Electric Current.—Its Chief Properties.—Heating Power.—Electrical Resistance.—Names of Electrical Units.—Chemical Power of an Electric Current.—Hydraulic Analogies.—Electric Pressure.—Fall of Electric Pressure down a Conductor.—Ohm's Law.—Joule's Law.—Units of Work and Power.—The Watt as a Unit of Power.—Incandescence of a Platinum Wire.—Spectroscopic Examination of a Heated Wire.—Visible and Invisible Radiation.—Luminous Efficiency.—Radiation from Bodies at Various Temperatures.—Efficiency of Various Sources of Light.—The Glow Lamp and Arc Lamp as Illuminants.—Colours and Wave-Lengths of Rays of Light.—Similar and Dissimilar Sources of Light.—Colour-distinguishing Power.—Causes of Colour.—Comparison of Brightness and Colour.—Principles of Photometry.—Limitations due to the Eye.—Luminosity and Candle-power.—Standards of Light.—Standards of Illumination.—The Candle-foot.—Comparison of Sunlight and Moonlight.—Comparison of Lights.—Ritchie's Wedge.—Rumford and Bunsen Photometers.—Comparison of Lights of Different Colours.—Spectro-photometers.—Results of Investigations.

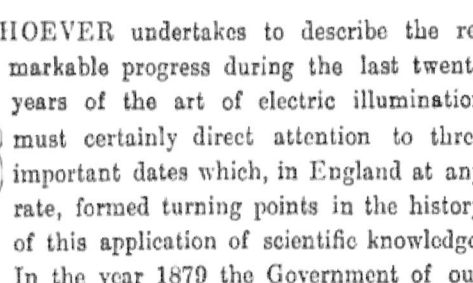

HOEVER undertakes to describe the remarkable progress during the last twenty years of the art of electric illumination must certainly direct attention to three important dates which, in England at any rate, formed turning points in the history of this application of scientific knowledge. In the year 1879 the Government of our country had its attention directed for the first time to electric lighting as a possible subject of legislation, and referred the whole matter to a Select Committee of the House of Commons, of which Lord Playfair was appointed Chairman. This Committee met, and took voluminous evidence from numerous and various experts, but the broad conclusion reached in the report which it finally presented was generally the expression of opinion that there was no reasonable scientific ground at that date for supposing that domestic electric lighting had obtained a sufficient footing to entitle it to be described as a practical success, and that, therefore, there seemed no useful result to be attained by interfering at once, by special legislation, with electric lighting. Passing over a gap of three years, we find in the year 1882 the whole aspect of affairs entirely changed by the completed invention of the electric glow lamp. At the beginning of that year the first Crystal Palace Electrical Exhibition enabled the public to properly appreciate the extent to which the then perfected electric incandescent lamp had revolutionised artificial lighting, and also the charm and beauty of that illuminant. Countless details, however, remained to be perfected, both in the glow lamp and in the devices for generating, distributing, using and measuring the electric current required for it. From the commercial point of view much information had to be slowly accumulated before even approximately correct opinions could be formed as to the revenue to be gained from the sale of electric energy for domestic purposes, and there was at that date but little information available for enabling an accurate forecast to be made concerning the probable average annual consumption of electrical energy by incandescent lamps when used instead of gas jets in different classes of buildings for illuminating purposes. There is no doubt, however, that the Electric Lighting Act of 1882, though much abused at the time, performed the important function of preventing the survival of immature schemes.

Between 1882 and 1888 exceedingly important improvements (some of which it will be necessary later on to examine) were made, and in 1888 the time seemed ripe for a fresh forward movement. This was effected by the pressure brought upon the Legislature to repeal one of the clauses in the Act of 1882, by which revision much more favourable conditions were created for inviting the support of capital; and as soon as the Electric Lighting Amendment Act of 1888 was an accomplished fact, a very important inquiry was held by the Board of Trade in May, 1889, in the Westminster Town Hall, London, under the chairmanship of Major Marindin. At this inquiry, which lasted for eighteen days, the whole subject of electric lighting by public supply, especially with reference to the needs of London, was carefully debated by many of the leading scientific and legal experts, and, as a result, the Metropolis was divided up into certain areas of electric supply, and conditions were laid down under which the distribution of current might be undertaken either by Public Companies or by the Local Authorities. Up to the date of that inquiry the total amount of electric lighting in London or in the Provinces had been, comparatively speaking, very small. From and after that date it has advanced by leaps and bounds. The progress made in the use of the incandescent lamp as a means of artificial illumination is shown by the following figures, and graphically indicated in Fig. 1.

At the end of 1890 there were probably in use about 200,000 incandescent lamps in London, but at the end of 1898 rather over two-and-a-quarter million 8 c.p. incandescent lamps were in employment. In the Provinces in 1890 there was hardly any electric lighting worth mentioning, whereas at the end of 1898 the total number of lamps in use in connection with public electric supply stations was just under three millions, the grand total for London and the Provinces being 5,206,000 8 c.p. lamps or their equivalent. There are at the present time (July, 1899) thirty-six electric lighting undertakings in and around London, including London proper and Greater London. These are in the hands of either public Limited

Liability Companies or of the Local Authorities. In the Provinces there are, at the same date, completed, or being completed, 150 electric lighting stations in as many towns and cities. In these the electric supply is likewise given either by the Local Authority or by a Limited Company. Whereas, however, in London the majority of the lamps in use are supplied by Companies, in the Provinces it is the reverse. The Local Authorities, in the provincial towns to a

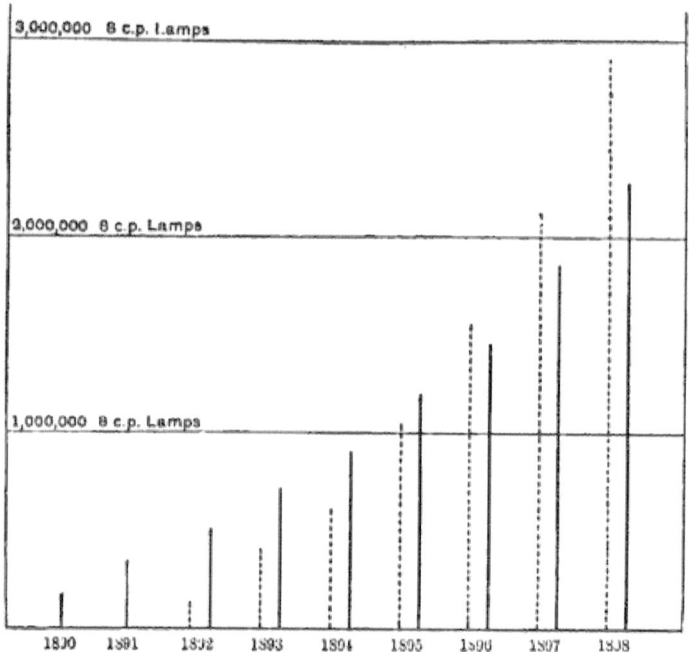

Fig. 1.—Diagram showing the Progress of Electric Lighting in London and the Provinces in Nine Years. The altitude of the vertical black lines represents to scale the number, from year to year, of 8 c.p. lamps installed. The firm lines are lamps in London and suburbs, and the dotted lines are lamps in the Provinces.

considerable extent, have assumed the position of "undertakers," as they are called in the Acts, but in a few cases have allowed the Provisional Order to be possessed by Public Companies. Hence in the Provinces of the United Kingdom electric lighting has largely become municipalised, and in many cities (such as Liverpool, Sheffield and others) where the public electric lighting has been originated by a Limited Company, the Corporation has ultimately purchased the undertaking, often for much more than its original cost.

In the ten years between 1888 and 1898 it may be said that the business side of public electric supply and the art of management and control of public electric generating stations became a specialised industry. The causes contributing to financial success or failure in this new industry were carefully studied, and, side by side with this attention to the commercial questions involved, steady progress in the mechanical and electrical improvements of the machinery of generation and distribution took place. Hence, at the close of the nineteenth century, we find a large number of new industries created by the demand for electric current for lighting. An assistance also has been given to pure science by the necessity for more exact knowledge on the electrical and magnetic qualities of the materials employed in the construction of electrical appliances. Collaterally, a strong impetus was given to the improvement of other mechanical machines, such as the steam engine. Engineers directed attention to the improvement of the dynamo in regard to its mechanical construction, and electricians provided new information, not only on the electrical theory of the dynamo, but on the principles affording a means for predetermining the results of operation for any particular design.

The result has been the evolution of a new machine—viz., the combined high-speed steam engine and dynamo—which places in our possession a most perfect and highly-efficient device for converting the energy of high-pressure steam into electric energy in the form of an electric current; the total loss in conversion not exceeding 10 per cent. in large plants.

Simultaneously with these improvements in the generating appliances proceeded others for effecting the distribution and use of electric currents, particularly in incandescent and arc lamps. Moreover, it was soon found that incandescent electric lighting lent itself very readily to decorative purposes in a manner impossible for the older illuminants. Hence the subject of "Electric Lamps and Electric Lighting" has not only a scientific and commercial, but also an artistic side, and has considerable practical interest for every householder.

Accordingly, in the following short treatise attention will be directed to the elucidation of facts which ought to be known by every user of the electric light, and the knowledge of which will enable him to understand something of the principles which underlie the art of electric illumination, and to comprehend as well the aid which it can render, when properly applied, to beautify and please.

It will be necessary to open the whole of our discussion by some simple illustrations of the meaning of fundamental terms. Every science as well as every art has its necessary technical terms, and even if these words at first sound strangely, they are not therefore necessarily difficult to understand. We are all familiar with the fact that electric illumination depends upon the utilisation of something which we call the electric current. Little by little scientific research may open up a pathway towards a fuller understanding of the true nature of an electric current, but at the present moment all that we are able to say of it is, that we know what it can do, how it is produced, and the manner in which it can properly be measured. Two principal facts connected with it are, that when a conductor, such as a metallic wire or a carbon filament, or any other material which is capable of being employed as a conductor, is traversed by an electric current, heat is generated in the conductor, and the space round the conductor becomes capable of influencing a magnetic needle. These facts can be simply illustrated by passing an electric current through an iron wire (*see* Fig. 2). You will notice that as the current is gradually increased the iron wire is brought up from a condition in which it is only slightly warm to one in which it becomes visibly red hot in the dark, and finally brilliantly incandescent. At the same time if we explore the region round about the wire with a suspended compass needle, we find that at every point in the neighbourhood of the wire the magnetic needle places itself, or tries to place itself, in a position perpendicular to the

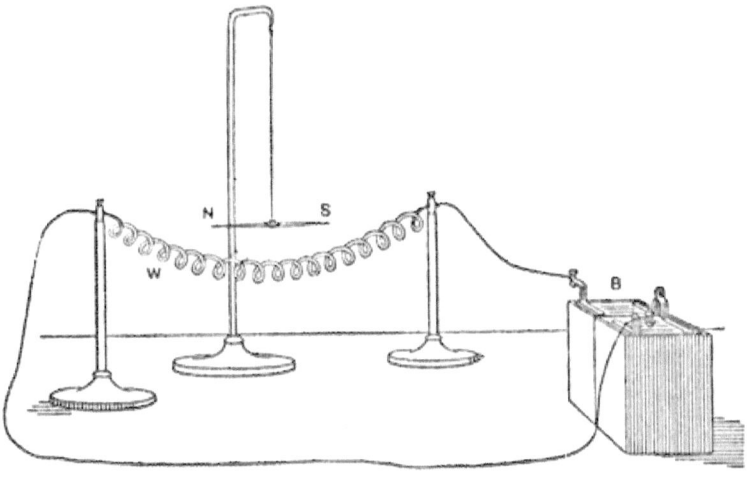

Fig. 2.—An iron wire **W** is rendered incandescent by an electric current sent through it from a battery **B**. The magnetic needle **N S** held near it sets itself across the wire.

wire. This fact, of capital importance, was discovered by H. C. Oersted in 1820, and in the Latin memoir in which he describes his epoch-making discovery he employs the following striking phrase to express the behaviour of a magnetic needle to the wire conveying the current. He says, "The electric conflict performs circles round the wire." That state which he called the electric conflict round the wire, we now in more modern language call the *magnetic field* embracing the conductor. We shall return in a later lecture to this last fact.

Meanwhile I wish at present to fasten your attention on the heating qualities of an electric current, and the laws of that heat production and radiation. The same electric current produces heat at very different rates in different conductors, and the quality of a body in virtue of which the electric current produces heat in passing through it is called its electric resistance. If the same electric current is passed through conducting wires of similar dimensions, but of different materials, it produces in them different quantities of heat in the same time. Before you (Fig. 3) is a chain composed of spirals of iron and copper wire. These wires are each of the same length and of the same diameter. Sending through this com-

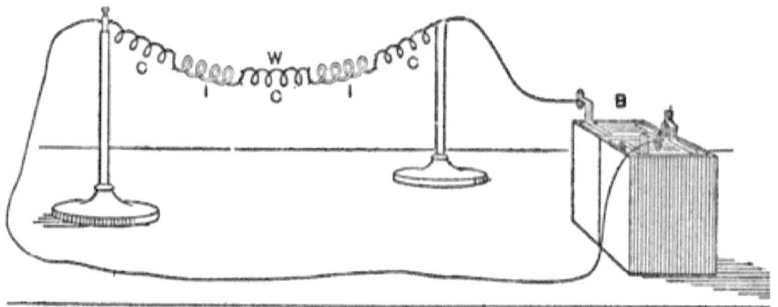

Fig. 3.—A chain W composed of alternate spirals of copper C and iron I is traversed by an electric current sent through it by a battery B. The iron links become red hot, the copper links only slightly warm.

pound chain an electric current, we notice that the iron wire links are very soon brought up to a bright red heat, whilst the copper links, though slightly warm, are not visibly hot. We have, therefore, before us an illustration of the fact that a current heats the conductor, but that each conductor has a specific quality called its electrical resistance, in virtue of which the same strength of current produces heat in it at a rate depending on the nature of the material. Other things being equal, the bodies which are most heated are said to have the highest resistance.

It is now necessary to notice the units in which these two quantities, namely, electric current and electric resistance, are measured. For the sake of distinction, units of electric quantities are named after distinguished men. We follow a similar custom in some respects in common life, as when we speak of a "Gladstone" bag or a "Hansom" cab, and abbreviate these terms into a gladstone and a hansom. Primarily the distinctive words here used are the names of persons, but by application and abbreviation they become the names of things. An electric current is measured in terms of a unit current which is called an *ampere*, and electrical resistance is measured in terms of a unit which is called an *ohm*, these being respectively named after two great investi-

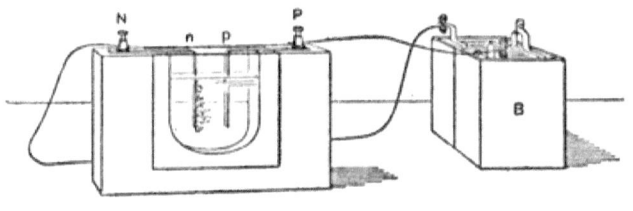

FIG. 4.—Glass lantern trough containing two lead plates n p, and a solution of sugar of lead. When a current from a battery B is sent through the cell it deposits the lead in tufts on the negative plate n.

gators, André Marie Ampère and Georg Simon Ohm. In order to understand the mode in which an electric current can thus be defined, we must direct attention to another property of electric currents, namely, their power of decomposing solutions of metallic salts. You are all familiar with the substance which is called sugar of lead, or, in chemical language, acetate of lead. Placing in a small glass trough a solution of acetate of lead and two lead plates (*see* Fig. 4), I place the cell in the electric lantern and project the image upon the screen. If an electric current is passed through the solution from one lead plate to the other it decomposes the solution of acetate of lead, extricating from the solution molecules of lead and depositing them on one of the lead plates, and you see the tufted crystals of lead being built up in frond-like form on the negative pole in the cell. We might employ, in preference to a solution of acetate of lead, a solution of nitrate of silver, which is the basis of most marking inks, and the same effect would be seen. It was definitely proved by Faraday that we might define the strength of an electric current by the amount of metal which it extricates from the solution of a metallic salt in one second, minute, or hour. The Board of Trade Committee on Electrical Standards have now given a definition of what is to be understood by an electric current of *one ampere* in the following terms: An electric current of one ampere is a current which will in one hour extricate from a solution of nitrate of silver 4·025 grammes of silver.* Otherwise we might put it in this manner: A current of electricity is said to have a strength of one ampere if, when passed through a solution of nitrate of silver, it decomposes it and deposits on the negative plate one ounce of silver in very nearly seven hours. We are acquainted in the laboratory with currents of electricity so small that they would take 100,000 years of continuous action to deposit one ounce of silver, and we are familiar in electric lighting practice with currents great enough to deposit one hundredweight of silver in thirty minutes. The simple experiment just shown is the basis of the whole art of electro-plating. Hence, when we speak later of a current of one ampere, or ten amperes you will be able to realise in thought precisely what such a current is able to achieve in chemical decomposition. It may be convenient at this stage to bring to your notice the fact that an 8 candle-power incandescent lamp working at 100 volts usually takes a current of about one-third of an ampere, a current which would deposit by electro-plating action one ounce of silver in about twenty-one hours.

---

* 20·3495 grammes = 1 ounce avoirdupois.

We pass next to consider another important matter, viz., that of electric pressure or potential; and we shall be helped in grasping this idea by considering the corresponding conception in the case of the flow of fluids. When a fluid such as water flows along a pipe it does so in virtue of the fact that there is a difference of pressure between different points in the pipe, and the water flows in the pipe from the place where the pressure is greatest to the place where the pressure is least. On the table before you is a horizontal pipe (Fig. 5) which is connected with a cistern of water, and which delivers

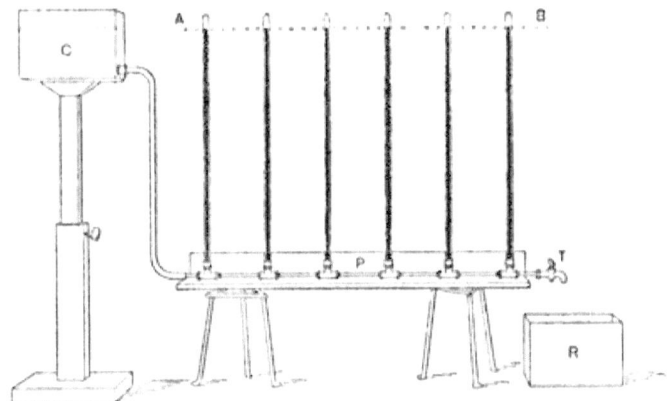

FIG. 5.—Horizontal pipe P, having six vertical gauge tubes attached to it. The pipe P is in connection with a water cistern C, and when the tap T is shut the water stands up at the same level, A B, in all the tubes.

water to another receptacle at a lower level. In that pipe are placed a number of vertical glass tubes to enable us to measure the pressure in the pipe at any instant. The pressure at the foot of each gauge glass is exactly measured by the *head* or elevation of the water in the vertical gauge glass, and at the present moment, when the outlet from the horizontal pipe is closed, you will notice that the water in all the gauge glasses stands up to the same height as the water in the cistern. In other words, the pressure in the pipe is everywhere the same.

Opening the outlet tap we allow the water to flow out from the pipe, and you will then observe that the water sinks (*see* Fig. 6) in each gauge glass, and, so far from being now uniform in height, there is seen to be a regular fall in pressure along the pipe, the gauge glass nearest the cistern showing the greatest pressure, the next one less, the next one less still, and so on, the pressure in the horizontal pipe gradually diminishing as we proceed along towards the tap by which the water is flowing out. This fall in pressure along the pipe takes place in every gas and water pipe, and is called the

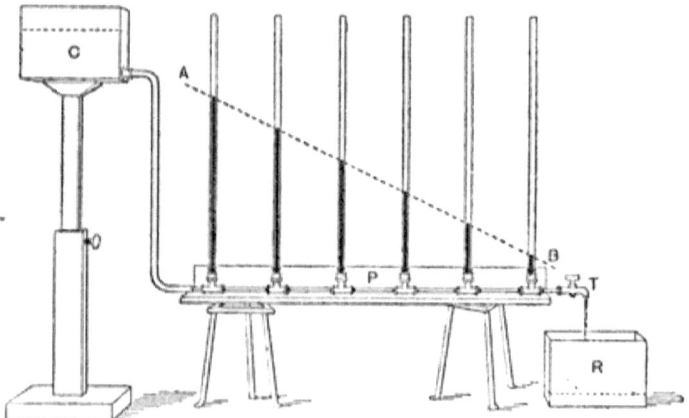

Fig. 6.—Horizontal pipe **P**, through which water is flowing from a cistern **C** to a reservoir **R**. When the tap **T** is open the water stands at gradually decreasing heights in the pressure tubes. The dotted line **A B** shows the hydraulic gradient.

hydraulic gradient in the pipe. The flow of water takes place in virtue of this gradient of pressure. It will be next necessary to explain to you that there is an exactly similar phenomenon in the case of an electric current in a wire, and that there is a quantity which we may call the *electric pressure*, which diminishes in amount as we proceed along the wire when the current is flowing in it. In order to understand the manner in which this electric pressure can be

ELECTRIC MEASUREMENT. 13

measured, a few preliminary experiments will be essential. Every body which is charged with electricity has, in virtue of that charge, a certain electrical potential, or pressure, as it is called, and electricity always tends to flow from places of

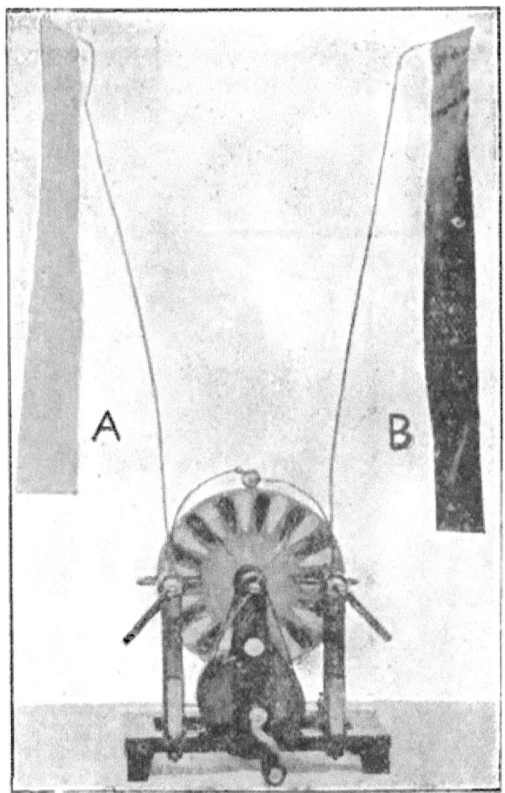

FIG. 7.—A small Wimshurst electrical machine, having two strips of paper A and B attached to its terminals. When the machine is worked the paper strips are drawn together.

higher to lower potential, just as water or other fluids tend to flow from places of greater to less pressure. When two bodies are at different electric pressures, or potentials, it is found that there is an attraction or stress existing between them, and a tendency for them to move, if possible, nearer together. If I attach to the terminals of a small electrical machine two paper strips, and then charge those paper strips to different electric pressures, we find the strips

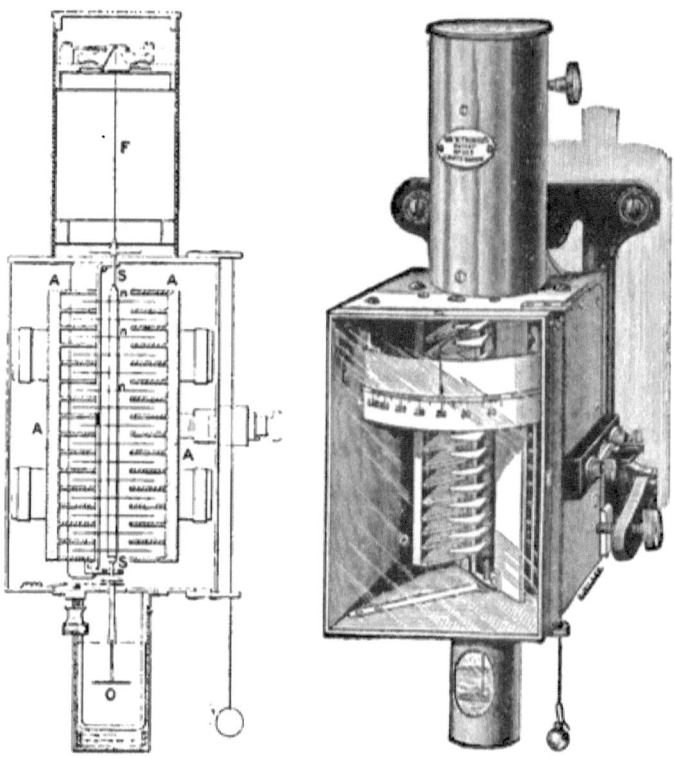

Fig. 8.—Lord Kelvin's Multicellular Electrostatic Voltmeter.
The fixed plates or cells are marked A in the sectional drawing. The movable plates n are attached to an axis S suspended by a fine wire F.

are drawn together (see Fig. 7). The difference of pressure between these two bodies can be exactly measured by the mechanical force with which they attract one another, or by the force required to keep them apart by a certain distance. This fact is taken advantage of to construct many instruments, which are called electric pressure-measuring instruments, or voltmeters. One of the most valuable of these is the electrostatic voltmeter, invented by Lord Kelvin. It consists (*see* Fig. 8) of a series of fixed plates, which are called cells, and suspended between these are a number of movable plates, all attached to a common axis, this axis being suspended by a very fine wire. The suspended plates are so arranged that, when at a different electric pressure or potential from the fixed plates, they are attracted in between them, and the movement of the suspended plates is resisted by the torsional elasticity

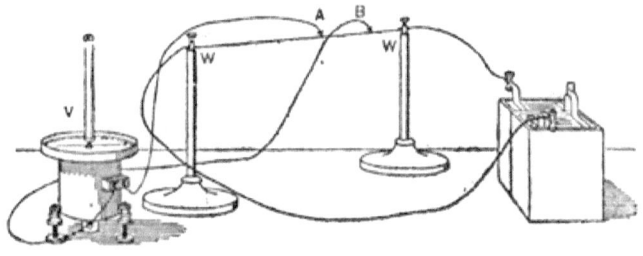

FIG. 9.—A battery C, which in the actual experiment consists of 50 cells, sends a current through a wire W W. By means of a pair of contact pieces A B, the terminals of an electrostatic voltmeter V are connected to various points on the wire W W, and the fall in electric pressure along it is thus measured.

of the suspending wire. The extent to which they thus move can be measured by an indicating needle fastened to the movable plates. If the fixed and movable plates in this instrument are brought to different electric pressures, or different electric potentials, they will be attracted towards one another, and we then have an instrument which can be converted by proper graduation into an electric pressure-measuring instrument. Furnished with such an appliance we can now explore the change in pressure down a wire through which the electric current is flowing. Through this manganese-steel wire we are now passing a current of electricity. One terminal of the voltmeter, namely, that connected to the fixed plates, is kept connected to one end of that manganese-steel wire, whilst the wire connected to the movable pieces of the voltmeter can be slid along the manganese-steel wire to different points (see Fig. 9). You will see from the indications of the voltmeter that as I slide the movable contact along the conductor conveying the current the indications of the voltmeter increase, and it is possible thus to show that there is a difference of electric pressure between two points on the wire conveying the current, precisely as there is a difference of fluid pressure between two points in a pipe along which water or gas is flowing. The water flows in the pipe from places where water pressure is greatest to places where water pressure is least. The electric current, whatever it may really be, likewise flows from places where the electric pressure is greatest to places where it is least. A unit of electric pressure has been selected which has been called a volt, after Volta, who, in 1800, gave us the first galvanic battery. It may be remarked in passing that the pressures at which it is most usual to work incandescent lamps are either 100 or 200 volts between the terminals of the lamp, and that the pressure between the terminals of a single galvanic cell, such as is used for working electric bells, is from 1 to 1½ volts.

We are then able to connect these two ideas of electric current and electric pressure when measured in the units above defined, and to give a definition of the unit in which electric resistance is measured. If we assume that a wire is taken through which an unvarying current of one ampere is passed always in one direction, and two points on this wire are found such that the difference of electric pressure between these points is one volt, then such wire is said to have a resistance of one ohm between those chosen points. The resistance of any conductor can be measured by comparing it by certain methods with that of a conductor whose resistance is one ohm, and the electrical resistance of any wire or conductor can thereby be expressed in units, each of which is called one ohm. The great service which Dr. G. S. Ohm rendered in 1827 to electrical science was that he gave the first clear definition of the manner in which electric current, electric pressure, and electric resistance are related to one another. This is now embodied in a statement which is a definition of the mode of measuring electrical resistance. If the strength of the electric current which flows in any conductor when that current has a uniform flow in one direction is measured in amperes, and the difference of the electric pressures between the ends of that conductor is given in volts, the electrical resistance of that conductor in ohms is obtained by dividing the last number by the first. The above statement as it stands is merely a verbal definition of what is meant by electrical resistance. It has, however, also been experimentally proved that the resistance of a conductor, as measured above, does not depend upon the value of the current, and is the same for large currents as for small ones if a correction is applied for the change of temperature of the conductor which is produced by the current. The resistance of a conductor is generally affected considerably by change of temperature; for some bodies, such as pure metals, it is increased, and for other bodies, such as carbon and certain metallic alloys, it is decreased.

Apart, however, from changes of temperature, it is an experimental fact that the strength of the current produced in any given conductor is exactly proportional to the fall in electric pressure down the conductor, and the numerical ratio of the numbers representing the fall of pressure in volts and the strength of the current in amperes is the value of the electrical resistance of that conductor in ohms. In order that you may see the application of this in electric lighting, let us consider the above in connection with an incandescent electric lamp. Such a glow lamp, as at present constructed, consists, as we shall see in the next lecture, of a fine carbon thread, or filament, which, as usually made, is traversed by a current of about two-thirds of an ampere, when that conductor is one which is suitable for a 16-candle-power lamp worked at the usual pressure of 100 volts. The electric supply companies bring into our houses two wires, between which they are constantly engaged in keeping an electric pressure difference of 100 volts or 200 volts, or thereabouts. If, therefore, the terminals of a lamp are connected to these two supply wires, the ends of the carbon filament are exposed to an electric pressure of 100 volts. The electric resistance of that carbon, when incandescent, is therefore expressed in ohms by dividing the number expressing the pressure difference in volts by the number defining the current in amperes; hence it is the quotient of 100 by two-thirds, or 150 ohms.

Another fundamental law in connection with the flow of an electric current in conductors was enunciated by Mr. Joule in 1841, and is called Joule's law. It is thus stated: If a current flows through an electric conductor, the heat produced in that conductor per second is proportional to the product of the square of the current strength as measured in amperes and the resistance of the conductor measured in ohms. Joule deduced this law from elaborately careful experiments made on the quantity of heat produced in a certain wire when traversed by an electric current, that wire being immersed in water. His experimental procedure was as follows:—He immersed a wire, formed into a spiral, in a vessel of water so protected as not to be able to lose heat from the outside. He then passed measured currents of electricity through the spiral, and observed with delicate thermometers the rise of temperature of the water in a stated time. The whole of the energy which is thus being spent in the wire is converted into heat, and that heat is employed in raising the temperature of the water. If by suitable means we prevent the loss of heat from the containing vessel, or otherwise take it into account, and if we try this experiment with currents of two different strengths, say of one ampere and two amperes, it will be found that, if the resistance of the conductor remains the same, the heat generated in a given time by the current of two amperes will be four times as great as the heat generated in the same time by a current of one ampere, and in like manner a current of three amperes would generate nine times as much heat as a current of one ampere, always provided that the resistance of the wire is not sensibly changed when the current is altered. A little consideration of the law of Joule and the law of Ohm, when taken together, will show you that, since the total amount of heat produced per second by a current of a given magnitude is proportional to the products of the numbers representing the resistance of the circuit in ohms, and the square of the strength of the current measured in amperes flowing through it; and, since the product of the value of the resistance of the circuit in ohms and the current strength in amperes is numerically equal to the difference of pressure between the two ends of the conductor, it follows that the total rate at which energy is being expended in any conductor to produce heat when a current of electricity is flowing through it is measured by the numerical product of the strength of that current in amperes, and the pressure difference between the terminals of that conductor measured in volts. If we apply this rule to the case of an electric lamp we find that, in order to measure the total rate at which energy is being transformed into heat and light in an incandescent electric lamp, we have to measure, in the first place, the current passing through it in amperes and the pressure difference between the terminals of the lamp in volts, and the product then gives us, in certain units, which are called *watts*, the rate at which energy is being dissipated or converted into heat in the carbon filament. Thus, for example, if a lamp which takes two-thirds of an ampere is placed upon a circuit having a pressure difference of 100 volts, the product of 100 and two-thirds being 66, the lamp would be taking 66 *watts*, and this is the measure of the rate at which energy is being supplied to the lamp, and converted by it into light and heat.

It is important that you should possess a very clear conception of the exact meaning attached to the *watt* as a unit of power. If a weight of one pound is lifted one foot high, the amount of exertion or work required to raise this weight is, in engineering language, called *one foot-pound* of work. If 550 pounds, or nearly one-quarter of a ton, are so lifted one foot high, the work done is called 550 foot-pounds. Imagine this last exertion made in one second, and repeated every second; the *rate at which work is being done*, or exertion made, will be equal to that which is called *one horse-power*. An ordinary man might for some time keep on lifting 55 pounds, say the weight of a good-sized full portmanteau, one foot high per second, and he would then be doing one-tenth of a horse-power. The work done in lifting one pound about nine inches high, or more nearly 0·7373 of a foot, has been selected as a unit of work, and is called one *joule*, and, if this work is repeated every second, this *rate of doing work*, or making exertion, is called *one watt*. It is necessary to notice carefully that *a watt* is a unit rate of doing work and not a unit of work. If the pound weight is lifted against gravity nine inches high slowly or quickly the work done is the same, viz., one *joule*. If it is lifted nine inches high in one second this rate of doing work is called *one watt*. If it is lifted nine inches high in half a second, the rate of doing work is doubled, and is then *two watts*. Hence it will be seen that one watt is a unit of power, which is the $\frac{1}{746}$th part of a horse-power. The reason for choosing such a small unit is that it gives convenient numbers in which to measure the rate at which work is done in such energy transforming agents as incandescent lamps. When a larger unit is necessary, it is usual to employ the *kilowatt*, which is equal to 1,000 watts, as a unit of power, and it is obvious that a kilowatt is equal to about one and one-third of a horse-power. Accordingly, such a lamp as we have above assumed dissipates energy at a rate which would require one horse-power to maintain eleven or twelve lamps, and the rate at which physical energy has to be supplied to a 16-candle-power glow lamp to keep it at full candle-power is almost as great as the maximum rate at which a strong man can do work.

Having prepared the way for further explanations by these elementary definitions, we must now proceed to examine more carefully what happens when an electric current is sent through a conductor, and the temperature of that conductor is allowed to rise. Coming back to our initial experiment with the iron wire heated by an electric current, we must note that there are three ways in which the energy supplied to that wire is being dissipated. In the first place, it is dissipated by contact with the cold air around it. The air molecules are continually beating against the hot wire, coming up to it cool, taking energy from it to heat themselves, and going away warm, and this process of carrying off of heat by the air molecules is called *convection*. In the next place, the wire loses heat by conduction out at the ends ; the warm wire is held in cool metallic supports, which are so large and such good conductors that they are not sensibly heated by the current. Hence, part of the heat of the wire is carried away by them, and if you examine the red-hot wire you will find that it is cooler at the ends than it is in the middle, being not quite so brilliantly incandescent close up to the clamps as it is in the centre. Then, furthermore, the wire is losing energy by a process called radiation. It is imparting energy to the *ether*, and sending out waves into this ether which represent an energy transformation, some of these waves being of such a character that they can affect the retina of the eye, and constitute what we call light; but by far the larger quantity of that wave energy is not capable of affecting the retina of the eye, and is called the non-luminous radiation. The luminous radiation, as we shall see later, in this case is only about one or two per cent. of the total radiation. Let us follow, then, the changes that take place as any conductor, say a wire or thin carbon rod, is gradually heated to incandescence. Prof. Draper, as far back as 1847, carried out such a series of experiments with a platinum wire, which he heated up gradually by an electric current, measuring at each stage the total amount of energy which was thrown out from the wire in the form of heat and the amount of energy thrown out from the wire in the form of luminous radiation or light. Before you is a table of Draper's results. His figures are in certain arbitrary units.

Table of Draper's Experimental Results (in 1847) on the Incandescence of Platinum Wire.

Temp'rature of the Wire in Centigrade degrees.	The Heat given out by the Wire.	The Light radiated by the Wire.	Remarks.
525°C	...	practically nil	Just visible in the dark.
527°C	0·87 units	,,	......
590°C	1·10 ,,	,,	Spectrum visible to line E.
653°C	1 50 ,,	,,	,, ,, ,, F.
718°C	1 80 ,,	,,	,, ,, ,, F-G.
782°C	2·80 ,,	,,	,, ,, ,, G.
910°C	3·70 ,,	,,	Full spectrum visible.
1038°C	6·80 ,,	0·39 units	......
1100°C	8 60 ,,	0·62 ,,	......
1166°C	10·0 ,,	1 73 ,,	Spectrum to H.
1230°C	12·5 ,,	2 92 ,,	......
1293°C	15 5 ,,	4·40 ,,	......
1367°C	...	7·24 ,,	......
1421°C	...	12·34 ,,	......

The above amounts of heat and light are given in arbitrary units. The letters E, F, G, refer to the fixed lines in the solar spectrum.

Let us, however, follow the whole progress of the phenomena experimentally by the employment of an incandescent lamp. Before me on the table is a carbon glow lamp, having a straight filament of carbon as the conductor to be rendered incandescent. By means of a lens we can project an image of the glowing carbon upon a screen, and by means of a prism we can expand that linear optical image into a prismatic spectrum, or rainbow strip, and notice the gradual changes which occur as the carbon is heated up from its lowest temperature to the highest incandescence it will safely bear. By passing a carefully graduated current through the lamp, we find out that at a certain point the carbon just begins to be visible in the dark. It has sometimes been inferred, as a deduction from Draper's experiments, and was, indeed, stated by him, that all bodies begin to be visibly red hot in the dark at the same temperature, namely, at about 525°C. This, however, is certainly not the fact. It has been shown by Weber, Bottomley, and others, as the result of careful experiments on carbon filaments and platinum wires, that bodies with a black, sooty surface have to be brought up to a higher temperature than bodies with a bright metallic surface before they begin to be visible in complete darkness. The crucial experiment has been made by J. T. Bottomley of taking two platinum wires, one of which is made to have a dull, sooty surface by coating it with the finest possible coating of lampblack, the other being highly polished. If these are placed in closed glass tubes from which the air is exhausted, and electric currents passed through the wires so as to heat them, it will be found, on carefully increasing the current and examining the wires in complete darkness, that the wire which has a bright metallic surface becomes visible in the dark at a lower temperature than the wire which has a dull, sooty surface. In Mr. Bottomley's experiments the temperature of the two wires was obtained from their electrical resistance; this last having been carefully measured at various temperatures previously. This and other experiments show that it is not strictly true that all bodies become luminous in the dark at the same temperature. But approximately we may say that most bodies, such as metals and carbons, when heated to 600° Centigrade, begin to give out radiation which is capable of affecting the eye as dull red light. Returning to our lamp, let us gradually increase the current through the carbon conductor of this Bernstein lamp, whilst at the same time we project the image of the straight carbon upon the screen, and examine it with a prism. As the current is gradually increased the carbon gives off radiation which is first entirely non-luminous, and which, though not capable of affecting our eyes, can be detected by a very delicate thermometer or other suitable instrumental means. As the temperature of the wire is increased to a higher point, radiation makes its appearance which is capable of affecting the retina of the eye. It is sometimes stated that the first colour which makes its appearance is red, but careful observation shows that the light which impresses the eye first, and which most easily stimulates the retina, is a *greyish-green*, which occupies in the spectrum the position of maximum brightness. If the temperature of the conductor is still further increased, we see that the prism shows us that red and yellow rays have made their appearance also, and a short spectrum is projected upon the screen in which the red, yellow, and green rays are clearly visible. Increasing still more the temperature of the conductor by passing a stronger current through it, we notice that the spectrum lengthens, greenish-blue, blue, and violet rays successively making their appearance, and are added to the others, whilst the previously existing green and red rays, and especially the green and yellow, are much strengthened. The spectrum, therefore, exhibits a growth, which growth consists in the addition of more and more refrangible rays, or rays towards the violet end of the spectrum, whilst at the same time a gradual increase in the intensity or luminosity of all the rays takes place, so that, when finally we have the complete spectrum exhibited on the screen, the conductor is found on examination to be in a state of brilliant incandescence, throwing out white light. We can, therefore, by the assistance of the prism, watch the gradual progress in the emission of radiation from the carbon conductor, which is capable of affecting the eyes; but experiments can be made which, at the same time, show that non-luminous or invisible radiation is being sent out from the conductor, and that this, at all stages of the increase in luminosity of the visible spectrum, is also increased in a certain proportion. Broadly speaking, therefore, as we increase the temperature of any body, it first begins by emitting rays which are not capable of affecting the eye, but finally, as the temperature is carried up to a higher and higher point, it emits successively rays which are capable of affecting the optic nerve; and in the end, when a temperature of 1,600 or 1,700°C. is reached, we have a total radiation from the conductor, which affects the eye as white light. The annexed table shows approximately the character of the light emitted from bodies when raised to different temperatures:—

Radiation from Bodies at different Temperatures.

	Centigrade.
Bodies begin to be just visible in the dark at about	400°
Dull red heat at	700°
Dull cherry-red heat at	800°
Full cherry-red at	900°
Melting point of silver at	945°
Clear red heat at	1,000°
Melting point of gold at	1,045°
White cast-iron melts at	1,040°
Orange-red heat at	1,100°
Bright orange heat at	1,200°
White heat at	1,300°
Steel melts at	1,300°
Bright white heat at	1,400°
Dazzling white heat at	1,500°
Palladium melts at	1,500°
Wrought-iron melts at	1,600°
Platinum melts at	1,775°
Iridium melts at	1,950°

Returning again to our incandescent wire, it must be noted that, of the total radiation sent out by such a hot body, only a small fraction of that radiant energy is capable of affecting the eye, and making its impression upon us as light. By far the larger proportion of radiation from most incandescent bodies is non-luminous radiation, and makes no impression at all upon the organ of sight. The proportion of luminous to total radiation from various sources has been measured by different observers, with the following results:—

Proportion of Luminous to Non-Luminous Radiation in Various Sources of Light.

Source.	Luminous Radiation.	Non-Luminous Radiation.
Red-hot wire	practically nil.	100 per cent.
Hydrogen flame	ditto.	100 ,,
Oil flame	3 per cent.	97 ,,
Gas flame	4 ,,	96 ,,
White-hot wire	4·5 ,,	95·5 ,,
Electric glow lamp	3 to 7 ,,	95 ,,
Arc lamp	5 to 15 ,,	90 ,,
Sunlight	34 ,,	66 ,,

The above table shows us, then, the percentage of luminous to non-luminous radiation in these various sources of light. The *luminous efficiency* of any illuminant is defined as the fraction, expressed as a percentage, which the luminous radiation is of the total radiation. We may also express, in the unit previously defined, called the watt, the rate at which energy is being expended in any illuminant to produce an illuminating power equal to that of one candle, and coupling this with the known luminous efficiency of the illuminant, we obtain two numbers which precisely express the energy consumption of the agent, and that which may be called its efficiency as a translating device for converting energy from one form, whether electrical or chemical, into another form— namely, eye-affecting wave-motion in the ether. Below are collected together these numbers for various sources of light:—

Efficiency of Various Sources of Light.

Source of Light.	Power Consumption in Watts required to produce a light of one Candle.	Ratio of Luminous to total Radiation, or Luminous Efficiency.
Candle	86 watts	2 to 3 per cent.
Oil lamp	57 ,,	3 ,,
Petroleum lamp	42·8 ,,	3 ,,
Argand gas lamp	68·8 ,,	4 ,,
Electric glow lamp	3½ ,,	3 to 7 ,,
Electric arc	0·8 ,,	5 to 15 ,,
Magnesium wire	......	15 ,,
Electric discharge in rarefied gases	......	33 ,,

One great problem awaiting solution in the future is the discovery of a source of artificial illumination which is, relatively speaking, much more efficient than those which are at present available; that is to say, of a method which will convert a much greater proportion of the energy transformed, into radiation strictly limited to that which can affect the eye; whether that energy be electrical energy in the form of an electric current, or whether it be chemical potential energy associated with materials to be combined. It appears probable that the solution of this problem rests in the development of phosphorescence into a practical method for yielding light. By the employment of apparatus of extraordinary delicacy, Prof. S. P. Langley and Mr. Very, in America, have succeeded in determining the luminous efficiency of the light emitted by the Cuban fire-fly. They find that in this natural light the whole energy of radiation is comprised within the limits of the visible spectrum, and, what is more important, chiefly within the limits of the green and greenish-yellow rays, which are especially important for the purposes of vision. These experiments show that the light emitted by this insect is, indeed, light without heat, and that its luminous efficiency is not far below 100 per cent. In our artificial production of light we have much lee-way to make up before we can rival the efficiency of the glow-worm and fire-fly.

Before, however, we can discuss this matter further, it will be necessary to turn our attention a little more at length to the subject of photometry, or the comparison of different sources of illumination in regard to their light-giving qualities. This is a part of our subject which, we may say at once, is in a much less satisfactory condition, as regards exact measurement, than the other practical electrical measurements to which reference has been made. The various light-giving bodies which we know and use, such as the sun, candle, gas lamp, incandescent electric lamp, electric arc lamp, &c., are bodies which are at very different temperatures, and they emit light, therefore, of very different composition. Every ray of light which has a pure and simple ray is characterised by two qualities—first, its *wave length*, which determines the *colour* impression it makes on the organ of vision; and, secondly, the *amplitude* of that wave motion, which determines its luminous *intensity*.

It may fairly be assumed that the reader is familiar with the fact that all optical research has indicated the fact that what we call light is a disturbance of some kind taking place in a universal medium called the ether. We do not know precisely the nature of that medium or that disturbance, but crucial experiment shows that, when a ray of light is passing through space, some change is being repeated very rapidly at every point in its path, and that this disturbance travels out from any point with a velocity which is, approximately, 186,000 miles a second; that is, it moves nearly one foot in the thousand-millionth part of a second. This disturbance is a wave motion—that is to say, at points in space separated by certain intervals similar motions or actions are taking place in a periodic manner at the same time, and the distance between two points in space at which similar actions or movements are taking place is called a *wave length*. We are familiar with such wave motion in the case of sound and the surface waves of water. By optical processes the wave lengths of light can be measured, and they have the values given in the table below. The unit of length in which these wave lengths are expressed is the one-hundred-millionth part of a centimetre.

Tables of Colours and Wave Lengths of Light.

Colour of the Light.	Wave Length.
Red	6,800
Orange-red	6,550
Orange	5,950
Yellow	5,680
Yellow-green	5,370
Pure green	5,200
Green-blue	4,910
Cyan-blue	4,700
Pure blue	4,580
Violet-blue	4,410
Violet	4,330

We may call to mind the fact, that in music the length of the air wave producing any note is exactly half the length of the wave corresponding to the note an octave below.

It is evident from the above figures that the eye is sensitive to about an *octave* of colour. It may assist the imagination if it is here mentioned that the wave length of green or greenish-blue light is about 1/50,000th part of an inch. This is not by any means an inconceivably small length. Gold leaf may be beaten out so fine that the thickness of a single sheet is only 1/300,000th part of an inch, or something like one-sixth part of the length of a wave length of green light.

In pure white light we may have rays of all these wave lengths present, together with any or all the intermediate ones whose wave lengths have not been given. Each different source of light emits in general a great group of rays of many different wave lengths, and each particular ray may be present in varying intensity or brightness. The proportion in which the different rays exist in any source, and the relative intensity which these respective rays have, is not the same for different sources of light. Hence the radiation from any illuminant is in general a very complex thing, and the effect which is produced when it falls on our eyes is a resultant or joint effect due to all the several individual rays that exist in that light. We have in the prism a means of analysing any compound ray by separating out the individual light rays which compose it in a fan-like form, so that we can detect which rays are present and which are absent, and measure also their relative brightness or luminosity. We did this just now when we expanded the linear image of our incandescent carbon filament into a broad many-coloured band called a spectrum. Two lights are said to be similar when their spectra can be made to have equal brightness in all their corresponding parts. That is to say, if the yellow ray in one spectrum is made equal in brightness to the yellow in the other, then the green, blue and red in the two spectra are also of equal brightness. A very little investigation shows us, however, that different sources of light are not similar. I throw upon the screen the spectrum of the light given by the carbon of an incandescent lamp, and also throw upon the same screen another spectrum just above the first, which is formed from the light of an electric arc lamp. These spectra are formed by exactly similar prisms placed symmetrically with regard to the screen, and you will notice that not only are similarly coloured rays in the two spectra not equally bright, but you will note the especial richness of the spectrum of the arc light in violet rays. We can so adjust these two spectra that they become exactly equal in brightness or luminosity for any particular ray, say the yellow—that is to say, we can so weaken the light of the arc lamp that its spectrum in the yellow is exactly equal in brightness to the spectrum of the incandescent lamp in the yellow. When this is done, we find that the arc lamp spectrum is very much brighter in the violet, green, and blue than that of the incandescent lamp spectrum, but that it is not so bright in the red. It is clear therefore, that these two lights are dissimilar in quality, and before we can speak of the comparison of two lights, we must understand distinctly what we are going to compare.

Broadly speaking, the two great purposes for which we require artificial light are to discriminate form and to discriminate colour difference. With regard to the discrimination of form, it is not necessary that the light which is employed should possess rays of more than one colour or wave length. Such a light is called mono-chromatic. If I illuminate this room with light which is wholly yellow or wholly red, by burning metallic sodium in the flame of a spirit lamp, or, in a similar manner, by burning some nitrate of strontium, which is the basis of red fire, we find that we are able to discriminate the form of surrounding objects, but that their normal colour differences have vanished. I need not spend much time in reminding you that what we call the colour of different objects is only a physical difference in their surface or substance which enables them to select certain rays out of a compound bundle of rays falling upon them for reflection, and to absorb the rest. Thus, a ripe cherry is red because, out of the whole collection of rays of light falling upon it, it sends back to our eyes by preference a large proportion of the red, yellow, and green rays, but absorbs some violet and blue; and the leaves of the cherry tree are green for the reason that they absorb a large proportion of rays other than green, but reflect more copiously these last. This selective reflection is, however, due to the absorption or diminution in intensity of certain rays in the compound light falling on the body, and which, passing *through* the surface of the reflecting body, is internally reflected, and then is returned back through the surface, having in the double transit suffered a deprivation or weakening of certain constituent rays. Hence, the colour of a body is dependent upon the character of the radiation which falls upon it, and if we illuminate a group of coloured bodies, first by light which is purely red, and then by light which is purely green, we find that in these two cases the apparent colour of the bodies is totally different.

These facts we are able to illustrate before you in a very simple manner by throwing some red light, produced by passing the light from the electric lantern through a pure ruby-red glass, upon a group of coloured ribbons or pieces of coloured paper. We note that the red paper still looks red in the red light, but that the green and blue paper or ribbons look almost perfectly black. On the other hand, when plunged into the green light, the red paper or ribbons now present an almost perfectly black appearance, whilst the green and greenish blue retain their normal colour. In order to understand in what sense sources of light can be compared together, it is necessary to keep in view the fact that every ray of light has three qualities: firstly, what we call its colour; secondly, its luminosity or brightness; and thirdly, its purity, or the degree to which that ray is mixed with white or other light.

When white light, the standard of comparison being daylight, falls upon a coloured surface, be it leaf or flower, some part of the light is reflected back unaltered in quality. Some portion passes through the surface, and is reflected back by internal reflection, and in performing the double journey through the surface layer it has certain of its constituent rays weakened or destroyed. The apparent hue or colour of the object depends upon the nature of this selective absorption. The brightness or luminosity of the surface depends upon the intensity of the light which falls upon it, and upon the amount by which that light which is reflected from it, whether superficially or internally, is weakened. The eye is sensitive to both these qualities of colour and brightness, and can pronounce judgment upon either of them, or on both. Moreover, we appreciate the extent to which the altered light is mixed with the unaltered light. In the case of coloured surfaces which we call *pale* or *light* in tint, such as a light pink, or light or Cambridge blue, there is a considerable admixture of white light when these surfaces are seen by normal daylight. Hence, when white light falls upon a surface, some of it is reflected unaltered, and some part is returned to the eye after having suffered a weakening or destruction of some, or all, of the rays present in it. A surface we call white reflects a large fraction of the light which falls upon it unaltered in quality; a surface we call black reflects a very small fraction, mostly unaltered, of the light which falls upon it; and a surface we call coloured sends back to our eyes a fraction of the light which falls upon it, but the quality of the light is altered by the diminution in brightness of some, or all, of the various rays present in it.

Suppose that there are two white surfaces, on each of which pure red light from different sources is being thrown. They may differ in brightness or luminosity, but there is generally no difficulty in deciding which of these two surfaces is the brighter. If, however, we have two surfaces, one of which is reflecting pure blue light and the other pure red light, then we have two differences to appreciate, namely, a real difference in colour due to the difference in the wave lengths of the lights reflected by these surfaces, and a difference in brightness or luminosity, which may, or may not, exist, due to the different intensities of the lights reflected from the two surfaces. A certain training of the eye is necessary in order to distinguish a difference in the luminosity of two surfaces, altogether apart from the question whether they have, or have not, a difference of tint. An inexperienced eye cannot do this with the exactitude of a trained eye. It is comparatively easy in extreme cases. If, for instance, I place before you, in a white light, a piece of dark blue paper and a piece of light yellow paper, you have no difficulty, quite apart from the colour difference, in deciding at once that the blue is darker than the yellow, or less bright; and if, on the other hand, I present to you similar pieces of a light Cambridge blue and a very dark red, you have also no difficulty in deciding as to the relative brightness of those surfaces. The artistic eye is especially trained in this sort of discrimination of brightness or luminosity, as distinguished from colour. An artist, for instance, does not admit that any objects are correctly depicted in which the proper differences of luminosity, or brightness, or of light and shade, as he would say, are not properly expressed or suggested. Very often it is quite out of the power of the artist to give the different parts of the surface of his paper or canvas the same actual relative luminosity or brightness as exists in the case of the objects he is depicting. He cannot, for instance, depict on his canvas dark green leaves when seen against a bright background of white cloud or snow in a manner which shall give them the proper relative brightness or luminosity as seen in Nature, especially when that paper or canvas has to be viewed in the interior of a room. A great part of the art of painting consists in suggesting differences of luminosity which cannot be exactly obtained. But when an artist makes a sketch with monochrome, or crayon, or pencil, he endeavours in some sense to so alter the surface of the paper in its various parts that, without regard to colour differences, these patches of the paper imitate more or less nearly the relative luminosity or brightness of the various parts of the surfaces of the objects to be depicted.

When, therefore, objects having what we call colour are viewed by the aid of two different illuminants, each sending out light of different qualities, as far as regards *the colour-distinguishing powers* of those two lights there is no sense in which they can be compared. As regards the power of distinguishing colours, we cannot compare an electric arc lamp with a candle, because no number of candles, however great, are, or can be, the equivalent of any arc lamp in their power of revealing or producing colour differences. We can, however, within certain limits, compare lights in regard to their different powers of producing on a white surface equal luminosity or brightness, whether these lights be monochromatic—that is to say, emit rays of only one wave length—or whether they are emitting a compound light composed of rays of many wave lengths. Taking any surface, such as white paper, which is capable of reflecting rays of all wave lengths, at least as far as those which affect the eye are concerned, we may illuminate part of this paper from one source of light and part from another source, and we may adjust the intensity of these two sources of light in such a way that a discriminating eye can assert that the two parts of the white surface are of equal brightness, whether they are apparently of the same colour or not. There are, however, certain great difficulties in doing this which must not be ignored when comparing together the illumination produced by two lights of different character.

It is necessary to bear in mind that in making these photometric comparisons, our eye is the only instrument which we can employ, and we are limited in our operations by the physiological properties of that organ. The eye is wonderfully susceptible of education, but there are certain inherent properties of it which no education can overcome or displace. It is often taken for granted that the physiological effect—namely, the amount of visual sensation produced by a given light—is proportional to the luminous or intrinsic brilliancy of that light, but this is not really the case. If two white surfaces are illuminated respectively by pure red and pure blue light, and the lights are adjusted so that the surfaces are of apparently equal brilliancy, then, if both the sources of light are doubled or trebled in intensity, the relative illuminations on those surfaces will no longer be equal. In other words, the visual sensation does not increase proportionately to the luminous intensity of the source of light. The visual impression produced by violet or blue light increases more slowly than that of red light when the absolute or intrinsic intensities of the two lights are steadily increased by equal steps. As long, however, as we are dealing with lights of fairly similar character, we have a practical basis for comparison in the experimental fact that the illumination produced on a white surface, due to any such source of light, say a standard lamp, placed at any distance from the surface, can be exactly imitated by placing four such equal sources of light at double the distance from the surface; provided that in both cases the rays of light fall in a similar manner on the surface.

It is owing to the above-mentioned different physiological action of different coloured lights (a phenomenon discovered by Purkinje) that high authorities such as Helmholtz ("Physiological Optics," p. 420) have stated that any comparison of lights of different colours is impossible. One fact is perfectly certain, and that is, that there is no way in which we can by any simple number, such as the candle-power, express the relative colour-distinguishing powers of different lights, and any attempt to do so is pure nonsense. Taking daylight from a bright northern sky as our standard of normal light, we cannot express the degree in which the light from an electric arc or glow lamp can reveal colour differences as seen by such normal daylight by stating the candle-power of those lights, and hence all such expressions as that an electric arc lamp has so many candle-power are insufficient and inaccurate methods of assigning a visual value to the light. We can, however, compare within certain limits the relative brightness of white surfaces when exposed to the two lights which are to be compared. For many purposes for which we require light, such as reading or writing, this gives us a measure of the value of the light for visual purposes. At present, the only scientific basis for photometry is to be found in the approximately correct fact that *brightnesses* can be added together, and that the illumination or brightness produced on a white surface by two separate lights is the sum of the brightness produced by the individual illuminations. The limitations set upon photometry by the physiological properties of the human eye seem, in some aspects, to be insurmountable; and, since *seeing* means discriminating colour and brightness difference, no substitute for the eye which is not an eye can be found. Our chief resource is the power we have of training the eye to increased visual sensibility, but a person with a naturally " bad eye " for colour or luminosity difference will never make a good photometrist.

In order to effect such a comparison of lights we must start, first, with a standard illuminant, and, second, with a standard of illumination. Unfortunately, the legal standard illuminant is agreed on all hands to be an exceedingly unsatisfactory standard. This standard is, by the Metropolis Gas Act of 1860, defined to be a standard sperm candle ⅞in. in diameter, burning 120 grains in the hour, and it is called a *standard or parliamentary candle*. We will return in a moment to consider the various other and much better standards of illumination which have been suggested, but, for the present, let us assume that by the candle we mean a normal standard candle, or, at any rate, the mean of a large number of such standard candles in illuminating power. The illumination which such a candle produces on a white surface, say a sheet of paper, held at a distance of one foot from it, is called *one candle-foot*, and the candle-foot is the unit of illumination, just as the candle is the unit of illuminating power. The principle on which practical photometry is based is the experimental fact that the brightness of the surface of white paper produced by two candles at a distance of one foot from the surface is twice that produced by one candle at a distance of one foot, and, moreover, that the illumination of such a surface produced by four candles at a distance of two feet is equal to the illumination produced by one candle at a distance of one foot. Similarly, nine candles at three feet and sixteen candles at four feet distance from the white surface produce the same illumination as does one candle at a distance of one foot. The candle-foot is found to be an exceedingly convenient illumination for the purposes of reading. Most persons can hardly read comfortably if the illumination on their book or paper is less than one candle-foot. The illumination in a well-lighted room may vary from two candle-feet on the tables to half a candle-foot on the walls and a quarter of a candle-foot on the floor, while a brilliant illumination, such as that required on a theatre stage, is from three to four candle-feet. The illumination in a street due to the street gas lamps may be from one-third to one-fourth of a candle-foot; in a picture gallery, from one to three candle-feet. The illumination due to full, high moonlight in London is only $\frac{1}{80}$th to $\frac{1}{100}$th of a candle-foot, depending on the clearness of the sky and altitude of the moon. These fractional figures must be understood as follows:—An illumination of a quarter of a candle-foot means illumination which would be produced by a light of one candle held at two feet distance from the white surface, and an illumination of $\frac{1}{100}$th of a candle-foot is the illumination produced by a candle held at a distance of ten feet from the white surface. The illumination of full sunlight may be from 7,000 to 10,000 candle-feet when the sun is high in a bright clear climate. The mean daylight in the interior of a well-lighted room may be from ten to forty candle-feet. Attempts have been made at various times to estimate the relative brightness of a white surface when held in full sunlight and when held in full moonlight. Various numbers have been obtained by different observers. Bouguer, in 1725, and Bond, in America, in 1851, made estimates of the relative brightness of sunlight and moonlight. Unless the altitudes of the sun and moon and atmospheric conditions are stated, these relative values do not mean very much. Roughly speaking, full sunlight produces an illumination from 400,000 to 700,000 times as great as that of full moonlight. It is difficult to read anything but fairly large type in full moonlight.*

---

* Bouguer dispersed the rays of the sun falling on a small hole by a concave lens and compared the diffused sunlight with the light of a candle. Bond formed the image of the sun on a sphere of silvered glass, and compared it when so weakened by reflection with a standard light. Bouguer found that the intensity of full sunlight was 300,000 times greater than that of full moonlight, and Bond found the ratio to be 470,980 to 1. More recently Prof. Young, in America, has made an estimate of the sun's mean candle-power. After correcting for atmospheric absorption, he finds the candle-power of the sun to be $1,575 \times 10^{24}$. Prof. L. Weber, of Breslau, finds that the red rays in sunlight are $1,110 \times 10^{24}$ times more intense than the corresponding rays in candle light, and the green rays in sunlight $2,294 \times 10^{24}$ times as intense as green candle-light rays. Since the sun's mean distance from the earth is 48×10^{10} feet, the sun's illumination at the earth is $\dfrac{1,575 \times 10^{24}}{48 \times 48 \times 10^{20}} = 7,000$ candle feet. Average moonlight is $\frac{1}{70}$th of a candle-foot, and hence full sunlight illumination, on a white surface, is to full moonlight in the ratio of 490,000 to 1. (10^{24} stands for a billion times a billion.)

We turn, then, to consider some of the methods of photometric comparison which depend on the principle that one source of light is said to have an illuminating power equal to that of another when the brightness of a white surface held at the same distance from both these sources is the same. Taking, for instance, the normal candle as a standard, any other source of light is said to have four candle-power if it produces the same apparent brightness on a white surface held two feet from it as does a candle if held one foot from the same surface; such comparison being made wholly with regard to luminosity or brightness of the white surface, and no attention being given to the difference in colour in any

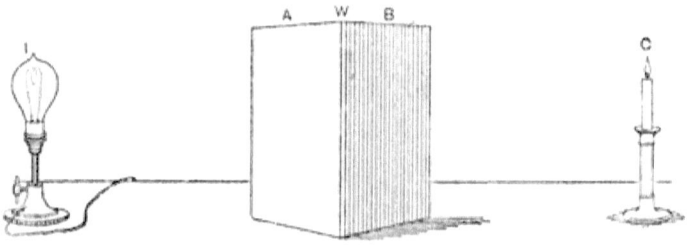

Fig. 10.—Comparison of Illuminating Power of Glow Lamp and Candle by means of Ritchie's Wedge.

white surface when viewed by the two sources of light. The simplest method of effecting this comparison is by the use of Ritchie's wedge. A block of wood is cut in the shape of a wedge with a rather obtuse angle, and has its two adjacent surfaces covered with fine white writing paper, the separating edge being made as sharp as possible (*see* Fig. 10). The two sources of light which are to be compared—say a candle and an incandescent lamp—are placed together in a darkened room. The wedge is held between them, with its edge vertical, and on looking at the vertical edge we see that one side of the wedge is illuminated by the candle and the other surface by the electric glow lamp. We can then move the wedge to and fro until we find a position such that the two sides of the wedge respectively illuminated by the two agents appear to be of the same brightness. We are assisted in getting rid of any difficulty due to difference of colour in the light by adopting the principle suggested by Capt. Abney, viz., by oscillating the wedge or swinging it to and fro. When we have very nearly found the position of balance, if we move the wedge to and fro in slowly diminishing arcs, we shall alternately increase and diminish the relative brightness of the two surfaces, but we shall not affect any difference in their relative colour tints; hence the attention of the eye is by this device fastened upon that which is varying, namely, the relative luminosity or brightness of the two surfaces, and distracted from that to which we wish to give no attention, namely, their relative colour difference. Having found the position in which the two inclined surfaces of the wedge appear equally bright when each respectively is illuminated by one of the two sources, we measure the distance from the candle to the wedge and from the lamp to the wedge. The illumination produced by the candle is then obtained by dividing unity, since the candle is the unit of illuminating power, by the square of its distance in feet from the wedge; and, likewise, the illumination produced by the electric glow lamp is equal to its unknown candle-power divided by the square of its distance in feet from the wedge. Since these illuminations are equal, it follows that the candle-power of the glow lamp must be numerically given by dividing the square of the distance of the glow lamp from the wedge by the square of the distance of the candle from the wedge. Thus, if the balance has been found when the glow lamp is four feet from the wedge and the candle is one foot from the wedge, the candle-power of the glow lamp is four times four divided by one times one, or 16.

Instead of employing a wedge, some observers make use of a method due to Rumford, in which the two illuminants are made to cast the shadow of a stick upon a white screen. If a glow lamp and a candle are placed a short distance apart (*see* Fig. 11), and a pencil or other rod of wood is held an inch or two from a white card, which is placed so as to be equally illuminated by the two sources of light, it will be found that there are two shadows on the card. One is a shadow due to the lamp, and the other is a shadow due to the candle. The shadow due to the candle is a space into which candle-light does not shine, and upon which the light from the lamp shines, whilst the shadow due to the lamp is a portion of the surface which is illuminated by candle-light alone. By suitably moving the

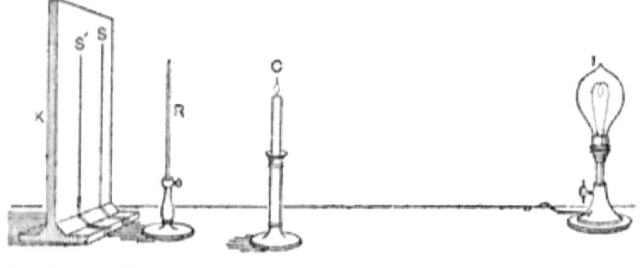

Fig. 11.—Comparison of Illuminating Power of Glow Lamp and Candle by means of Rumford Shadow Method.

candle and the lamp we find we can obtain positions in which the two shadows are of equal depth, and we then obtain the candle-power of the incandescent lamp by dividing the square of its distance in feet from the screen by the square of the distance in feet of the candle from the screen. All that we are doing in this case is to compare together the illumination produced upon a white surface which is partly illuminated from one source and partly from another, and the advantage which the rod gives us is that we can make these two shadows or illuminated surfaces just touch one another, and so be assisted in making a comparison between their relative brightnesses. If the same experiment is tried with moonlight and a candle, the observer will very soon realise the difficulties that attend photometric measurement when the lights are of different qualities. On any bright moonlight night hold a white piece of card at the window, and place a candle so as to illuminate the card at about the same angle. Hold a pencil in front of the card, and it will be seen that there are two shadows, one of which is a bright blue and the other a bright yellow. The yellow shadow is, apparently, the shadow thrown by the moon, and the blue shadow is, apparently, the shadow thrown by the candle. In reality, that which we take for the shadow due to the moon is a space illuminated by candle-light alone, and the shadow we take for the shadow produced by the candle is a space on the card illuminated by moonlight alone.* These two surfaces differ not only in colour, but in brightness, because the light falling upon them is very different in quality. By moving the candle we can find a position, generally from seven to ten feet from the card, in which the blue and yellow shadows, though different in colour, have apparently the same depth. But the difficulty of determining when this relative luminosity is the same will give an observer who tries the experiment for the first time an insight into the difficulties of photometric measurement.

A third and more frequently-used method for the comparison of sources of illumination is the grease-spot photometer of Bunsen. If a piece of thin paper has a spot of grease placed upon it, and if we hold up the paper between our eye and the light, we find the grease-spot more transparent than the rest of the paper, and it looks lighter. But if we hold the paper so that the light falls on it, we see that the grease-spot is apparently darker than the rest of the paper.

---

* It was this and other similar phenomena which drew Goethe into some confusion of thought on the subject of colour, and led him in his *Farbenlehre* to dispute Newton's theory of colour.

If the disc of paper with a grease-spot upon it (see Fig. 12) is held up between two lights, such as a gas flame and a candle, and looked at, first from one side and then from the other, it will be found possible, by moving the disc, to find a position for the piece of paper such that the grease-spot can hardly be seen from either side. When this is the case the two surfaces of the paper are equally illuminated by the two sources of light, and if one of these sources of light is a standard candle, then, as above explained, the candle-power of the other source of light is obtained by dividing the square of the distance in feet of the source of light from the screen by the square of the distance in feet of the candle from the screen.

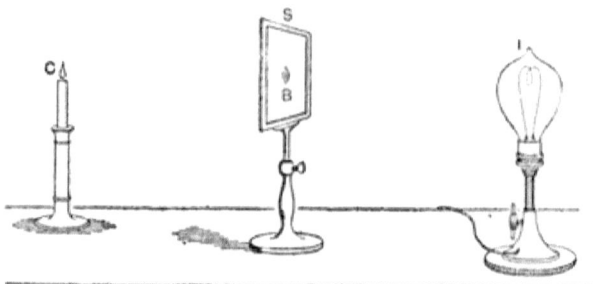

Fig. 12.—Comparison of Illuminating Power of Glow Lamp and Candle by means of Bunsen Grease-Spot Method.

In practice various devices are used, such as a pair of inclined mirrors at the back of the disc, and arrangements for reversing the disc, to enable the observer to rapidly make a fair comparison between the two surfaces of the disc, and to ascertain if, when equally viewed from both sides, the grease-spot is invisible. The grease-spot method is one which is largely used in the determination of the candle-power of illuminating gas. The Gas Companies are bound to supply gas of a certain candle-power when burnt from a particular standard burner, and official comparisons have to be made in order to determine if they comply with their obligations. If the grease-spot or shadow methods are employed for the comparison of a candle with a glow lamp, we do not find any serious difficulties in determining the relative brightness of the white surfaces illuminated by the two sources of light, because the qualities of these two lights are not very different—that is to say, if we formed a spectrum from each light we should find that the different rays in these spectra, which are of the same wave length, can all be made to be equal in luminosity; or, in other words, if the luminosity of the red ray in the glow lamp is equal to the luminosity of the red ray of the same wave length in the candle, then the green, blue and violet rays of corresponding wave lengths are also equal as regards luminosity. If the experiment is tried with two lights of very different qualities, such as an electric glow lamp and an electric arc lamp, considerable difficulty will be found by the unpractised observer in determining when two adjacent surfaces illuminated by these two sources of light are of equal brightness. In order to diminish this difficulty, observers have sometimes employed coloured glasses to select particular rays from the arc lamp and compare them with similar rays in the candle-light, and numerous determinations exist of the candle-power of different arc lamps for red and green rays. But this really serves no useful purpose. The practical value of a light is not the relative intensities of certain rays when compared with rays of the same wave length in the light of a candle, but the total brightness which that light is capable of producing on a white surface when compared with the total brightness which the unit illuminant is capable of producing on the same surface.

We may further illustrate by an experiment the difficulty of comparing together the brightness of two surfaces of different colour in the following manner:—An electric glow lamp with a red glass bulb and an electric glow lamp with a green glass bulb (*see* Fig. 13) are so placed as to equally illuminate a white surface. A rod or other opaque body is then placed so as to cast two shadows, one from each lamp. You observe that the two shadows are respectively red and green; the red lamp appears to cast a green shadow and the green lamp appears to cast a red shadow, and you will find considerably more difficulty in comparing together these two lights by this Rumford method of photometry than would be the case if the two lights were both of them white or both of them red. By adopting the method of oscillating one of the lights so as to

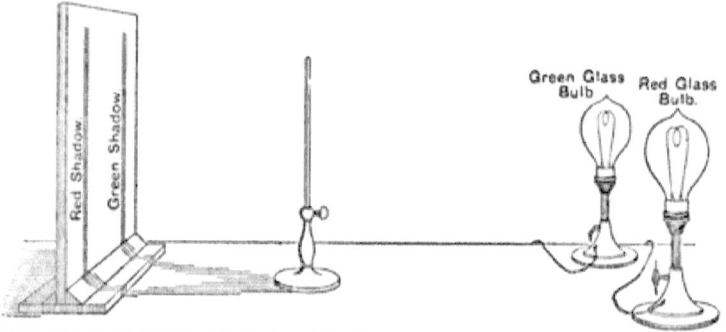

Fig. 13.—Comparison of Illuminating Power of Two Lights of Different Colours by Shadow Method. The two glow lamps have glass bulbs of different colours.

alternately increase and diminish the brightness of one of the shadows, the difficulty of making this luminosity comparison is partly overcome.

With regard to the standards of illuminating power, it has been previously stated that the British standard candle is an unsatisfactory standard. The French adopted a standard oil lamp, called a carcel standard lamp, burning 42 grammes of colza oil per hour with a wick of a certain size. This carcel standard gives a light equal to about $9\frac{1}{2}$ British candles. A standard which has been adopted to some extent in Germany, and to a less extent in England, is the Hefner-Alteneck amyl-acetate lamp. This is a little spirit lamp burning pure pear oil from a wick of certain size, and yielding a flame 40 millimetres high, with an illuminating power of rather less than one British candle. But the objection to its use as a standard is that it gives a somewhat reddish light. Another standard largely used in England is the Methven gas standard. In this standard, ordinary coal gas, enriched with benzol or pentane, two hydro-carbon liquids, is burnt at a particular form of argand burner. In front of the flame is placed a metal plate, having a slit in it of such a size that it only permits light to pass from the centre of the flame, and this slit is adjusted so that the light is equal to two standard candles. Variations of quality and pressure in the gas are said not to affect the intensity of the light which is emitted through the slit. In practice this is not found to be exactly the case. By far the most satisfactory standard which has yet been proposed is the pentane air-gas standard of Mr. Vernon Harcourt. In this standard a mixture of volatile hydro-carbon (called pentane) and air is burnt at a jet of a certain kind in such a manner as to yield a flame, produced by the combustion of a definite chemical compound, such flame having a certain height and dimensions. By taking suitable precautions, this pentane gas standard can be made to yield excellent results as a standard of light. It has been proposed by M. Violle, a distinguished French chemist, that an absolute standard of light should be obtained by taking the light emitted from one square centimetre of molten platinum. Such a standard, as a practical standard of reference, is not very convenient or suitable for use as a general primary standard of light, but it may be found valuable as an ultimate reference. For some purposes the only satisfactory method of comparing together two sources of light, and examining their qualities in regard to the relative proportions and brightness of the different rays existing in those lights, is by employing the spectro-photometer. This instrument consists of an arrangement whereby the same prism is made to analyse the light from two sources of light, and to expand the rays sent out from these two sources of light into two rainbow bands or spectra in such a manner that these spectra are placed one above the other, the corresponding colours being vertically over one another. It is then possible to weaken one of these sources of light by interposing either a disc with an aperture in it of variable size, revolving at a rapid rate, or by employing a pair of nicol prisms, by means of which the light from one source can be weakened to any extent. The brighter of these sources of light is weakened until one particular ray, say the yellow ray, in that light is exactly equal in brightness to the corresponding yellow ray in the spectrum of the other source of light. In other words, the spectra are made to match each other exactly at this particular spot. It will then be found that if the two sources of light are of different qualities, say a glow lamp and an arc lamp, the two spectra will not match each other as regards brightness in any other places.

Prof. Nichols, of Cornell University, U.S.A., has carried out a series of observations with an apparatus of this kind. Taking an Edison 16-candle-power incandescent lamp as his standard, he calls the brightness of the spectrum of this light everywhere unity, and he compares with it the spectrum of any other light, say an arc lamp, of which the spectrum has been weakened so as to make it of identical brightness for one particular yellow ray with the corresponding ray in the spectrum of the glow lamp. It is then found that the spectrum of the arc lamp is less bright in the red rays, but much more bright in the green, and especially much more bright in the violet. In the spectrum of the arc lamp, especially when that spectrum is taken in such a way as to utilise a large portion of the light proceeding from the carbon vapour between the poles, a very bright violet band is found at the extreme end of the spectrum. These observations have been expressed by Prof. Nichols in a series of curves (*see* Fig. 14). The horizontal line at the base of the diagram represents the spectrums; the letters A, B, D, E, &c., indicate

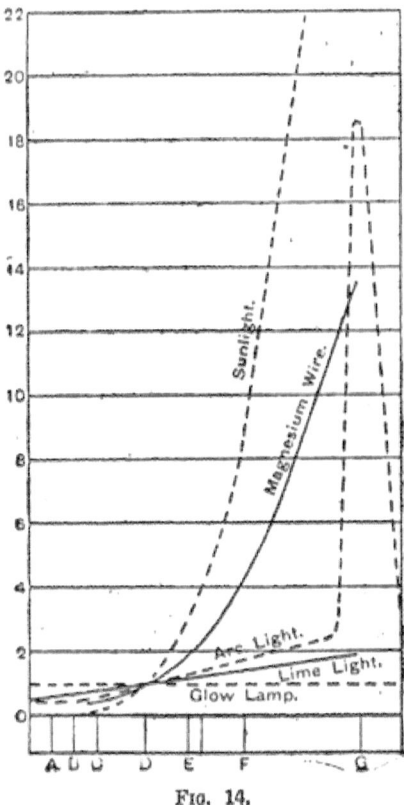

Fig. 14.

the position of the characteristic black lines or missing rays in the solar spectrum, which are taken as points of reference. The intensity of the spectrum of the glow lamp is taken everywhere as unity, and is represented by the altitude of the horizontal line marked *glow lamp*. The other curves then represent, by their ordinates or heights at various points, the relative brightness of the respective rays in the different portions of the spectra of the light taken from the sun, the arc lamp, and magnesium wire, compared with the ray of the same wave length in the light from the electric glow lamp. It will be seen how enormously brighter the sunlight and the arc light spectrum are in the neighbourhood of the violet rays when compared with the corresponding ray of the glow lamp, but that their light, relatively speaking, is less bright in the red than the corresponding light of the glow lamp, provided that the spectra of all these sources of light have been equalised so as to make them of identical brightness in the neighbourhood of the D line of the solar spectrum—that is, of the yellow rays.

LECTURE II.

The Evolution of the Incandescent Lamp.—The Nature of the Problem.—Necessary Conditions.—Allotropic Forms of Carbon.—The Modern Glow Lamp.—Processes for the Manufacture of the Filament.—Edison-Swan Lamps.—The Expansion of Carbon when Heated.—Various Forms of Glow Lamps.—Focus Lamps.—High Candle-power Lamps.—Velocity of Molecules of Gases.—Kinetic Theory of Gases.—Processes for the Production of High Vacua.—Necessity for a Vacuum.—Mean free path of Gaseous Molecules.—Voltage, Current and Candle-power of Lamps.—Watts per Candle-power.—Characteristic Curves of Lamps.—Life of Glow Lamps.—Molecular Shadows.—Blackening of Glow Lamps.—Self-recording Voltmeters.—Necessity for Constant Pressure of Supply.—Changes Produced in Lamps by Age.—Smashing Point.—Efficiency of Glow Lamps.—Statistics of Age.—Variation of Candle-power with Varying Voltage.—Cost of Incandescent Lighting.—Useful Life of Lamps.—Importance of Careful "Wiring."—Average Energy Consumption of Lamps in Various Places.—Load Factors.—Methods of Glow-Lamp Illumination for Production of Best Effects.—Artistic Electric Lighting.—Molecular Physics of the Glow-Lamp.—High-Voltage Lamps.—Varieties of Carbon.—Densities and Resistances.—Deterioration of Carbon Lamps.—High Efficiency Lamps.—Recently Suggested Improvements.

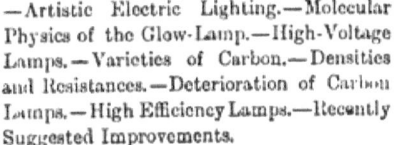

THE illustrations and explanations in the previous Lecture will have prepared the way for a study of the electric incandescent lamp as a source of illumination. We shall not occupy time by describing the stages by which the glow lamp has attained its present perfection. Neither is it necessary to dwell upon questions of scientific or legal priority which are not most advantageously discussed from the lecture table. Suffice it to say that in its modern form the carbon filament electric lamp has been, and always will be, inseparably associated with the names of Edison and Swan, and, in a lesser degree, with Lane-Fox, Sawyer and Man. These inventors were preceded by many who attempted solutions of the problem, and they have been followed by others who have added countless details of invention to the methods and means by which progress has been made from early ideas to a perfection which is even now not yet final. These successful workers, however, started from a standpoint similar to that which initiated the ideas and failures of early pioneers in this region of discovery and invention. The knowledge of the simple fact that a metallic wire or other substance could be heated by an electric current, and that such a heated body when raised to a proper temperature gave out rays of light, was the starting point for all the attempts made by early and by later inventors to discover a means of producing a conductor which would fulfil the following conditions:—(1) It must be capable of being heated to a high temperature without fusion or sensible volatilisation; (2) it must be a material of high specific resistance; and (3) it must be capable of being shaped into such a form as to be conveniently and economically utilised as a means of producing small units of light.

In its most widely used form the carbon filament glow lamp is based upon the fact that carbon can be heated in a highly perfect vacuum or in rarefied gases to a temperature near, or higher than, the melting point of platinum without suffering very rapid change, and that in certain forms it fulfils the three conditions named above, as necessary for the incandescing material in a glow lamp. Carbon is capable of existing in three modifications, which are chemically termed *allotropic* forms. These are: *charcoal* as ordinarily obtained by the carbonisation of some organic substance, such as wood, paper, thread, silk, and other similar substances, which consist principally of carbon united with other elements, which can be driven off by heat. Next, carbon exists in the form known as *graphite*, and we have it in this form in plumbago as used in lead pencils, and in the hard carbon deposit in gas retorts. And, lastly, carbon occurs in a transparent or opaque crystalline form known as *diamond*. All forms of carbon are by a sufficiently high temperature converted into graphite. Of these three forms of carbon, diamond is practically a non-conductor of electricity; ordinary charcoal or carbon, such as that produced by carbonising wood, cotton, or other organic materials, has an electrical conductivity which varies greatly with the circumstances under which it is produced; but, approximately speaking, it may be said that the hard and dense form of carbon which is obtained from gas retorts as a product of the distillation of coal, and which is very largely composed of graphitic carbon, has an electrical resistance which is, for the same volume, from one to three thousand times that of copper. The temperature of fusion and temperature of volatilisation of carbon is also very high. If carbon is heated to a temperature near to red heat in ordinary air, or in an atmosphere containing oxygen, an immediate combustion of the carbon takes place, produced by the union of the carbon with the oxygen of the surrounding atmosphere and the resulting production of carbonic acid gas. Hence, in order that permanence may be obtained, it is essential that the heating of the carbon should take place either in a vacuous space or else in an atmosphere which does not contain free oxygen. The practical invention of the incandescent lamp, therefore, turned upon the discovery of the proper method for producing carbon in the form of a thin wire, thread, or filament, as it is called, and supporting this carbon filament in a glass bulb quite free from all oxidising gases or air.

The final outcome of failure and research was ultimately to show that in order to obtain light by the incandescence of carbon, such incandescence being produced electrically, the following conditions must be met:—First, the carbon must be in the form of a flexible strip, wire or thread, technically termed a "filament," and must be in a dense conductive form produced by heating some organic substance to a very high temperature. An additional deposit of carbon is generally put upon the filament by a process called "treating." Secondly, the carbon must be enclosed in a bulb which is exhausted of its air as perfectly as possible. It has been frequently asserted by different inventors that carbon is equally or more permanent when contained in glass bulbs filled with rarefied gases, such as nitrogen or hydrogen, chlorine, bromine, or hydrocarbon gases; but it has not yet been definitely shown that these methods are as efficient as the employment of a very high vacuum. Thirdly, the enclosing bulb must be entirely of glass. Fourthly, the electric current must be got into and out of the conductor by means of platinum wires sealed through the glass. Without stopping to detail the attempts which have been made, or which are still being made to modify these conditions, or to construct lamps to give light by incandescence which do not fulfil them, suffice it to say that the electric glow lamps in general use (in 1899) all comprise the above four elements: carbon, vacuum, platinum, glass. Let us consider each of them in turn.

The carbon conductor or filament is made by carbonising some material which must be capable of being reduced to nearly pure carbon *after* it has been given the necessary filamentary form. This carbonisation is conducted by heating a suitable material reduced to a thread-like form to a very high temperature in a closed crucible. If a piece of paper, linen, cotton or thread, which consists essentially of cellulose (the chemical constitution of which is carbon united to oxygen and hydrogen), is raised to a high temperature, say a white heat, in a closed crucible made of very infusible material, such as compressed blacklead or plumbago, the hydrogen and oxygen are driven out from their combination with the carbon, and a residue is obtained which is more or less pure carbon. The temperature at which this carbonisation is conducted has a great effect upon the kind of carbon produced. Carbonisation at a very high temperature, near the melting point of steel, when conducted out of contact with air, produces a highly dense and elastic form of carbon. Sometimes the material employed for this purpose is cotton thread of a pure kind which has been *parchmentised* by being treated in a bath of sulphuric acid, commonly known as oil of vitriol, and water. This bath destroys the fibrous structure of the thread, and produces a material somewhat resembling catgut. Parchmentised paper is not infrequently used as a material for the covers of jam-pots.

The dilute sulphuric acid with which the thread is treated has the power of producing a change in the cellulose, which results in the formation of the above tough material. The parchmentised cotton thread is cut into the necessary lengths, wound on a frame made of carbon rods, and buried in a crucible which is packed full of powdered plumbago. The closed crucible is then exposed to a vivid white heat in a furnace, and after a sufficient exposure to this temperature the crucible is opened and the carbonised thread is taken out. It is then found that the threads have been converted into a material which is almost wholly pure carbon of a dense variety and highly elastic. In place of using parchmentised thread, as first suggested by Mr. J. W. Swan, other means have been adopted for preparing the cellulose in the form of a fine structureless thread. Pure cotton wool, which is nearly pure cellulose, is capable of being dissolved by several chemical materials, such as chloride of zinc, and the gummy mass so produced can be pressed out into a thread. This thread is then treated as above described, and converted into carbonised thread or carbon filaments of the necessary form. When so prepared the carbon thread is an elastic material, having an electrical resistance which is approximately two thousand five hundred times that of a copper wire of the same length and thickness. An immense variety of materials have been used and suggested for producing the carbon filament. At one time Edison used largely very thin slips of Japanese bamboo cut into the requisite shape. We will project upon the screen the image of an Edison lamp, of which the carbon filament is formed of carbonised bamboo and has been broken on

FIG. 15.—Modern Electric Glow Lamp (Edison-Swan), with Looped Carbon Filament.

one side, and set it in vibration; you will see by the rapid movement of the carbon loop when the lamp is shaken how exceedingly elastic is the delicate carbon filament. Spiral springs can even be prepared in this manner from carbonised thread which have a considerable degree of elasticity and tenacity.

The carbon thread, having been thus prepared, has next to be mounted in a glass bulb, and it must be so held that when heated the expansion which it experiences can be permitted to take place without endangering its rupture or severance from the wires to which its ends are fastened. Early experimentalists found an insuperable difficulty in connecting their carbons to the leading-in wires, and in constructing any practical lamp by the use of straight carbon rods. In most modern lamps the carbon conductor is given a horse-shoe or double-loop form, as shown in Fig. 15, in order that it may have perfect liberty to expand when heated without becoming detached from, or from putting any strain upon, the supporting wires by which it is held. It is very easy to show that carbon, like most other solid bodies, expands when heated. I have here an Edison-Swan incandescent lamp with a long straight carbon filament (*see* Fig. 16). One end of the carbon filament is attached to a platinum wire

FIG. 16.—Electric Glow Lamp with Straight Carbon Filament for showing the Expansion of the Carbon.

sealed through the glass at the top, and the other end is attached to a platinum wire sealed through the other end of the glass, but a little spiral spring of steel is inserted between the leading-in wire and the carbon. Projecting an image of this upon the screen, and asking you to fix your attention upon the end of the carbon attached to the spiral spring, it will be seen that when the electric current is sent through the carbon filament, making it incandescent, the spring contracts, thus showing the expansion of the carbon thread. If the carbon is rigid and is not given a loop or horse-shoe form, then it is necessary to make provision for this expansion of the carbon by attaching it to flexible supports or springs, as was done in the Bernstein lamp (*see* Fig. 17). The carbon filament, before inclusion in the bulb, is frequently subjected to a process which is called *treating*. If the carbon filament is rendered incandescent in an atmosphere of hydro-carbon gas—if, for instance, it is rendered incandescent when it is immersed in coal gas or benzol vapour—the heated carbon decomposes the hydro-carbon gas, and a deposit of carbon of a very dense form is made upon the filament, the carbon

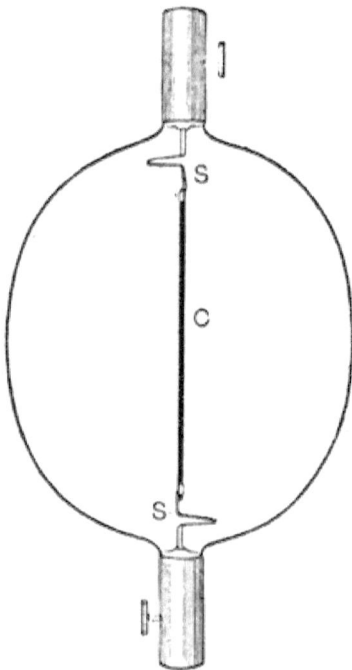

FIG. 17.—Bernstein Lamp, with Straight Carbon C, and Springs S to permit Expansion.

which is so deposited being of a graphitic form, and having, under the proper conditions of deposit, a brilliant steel-like lustre. This process is not, however, applied to all carbon filaments. This deposited carbon has a lower electrical resistance for the same volume, viz., about one-sixth or seventh, when compared with the carbonised cellulose of which the carbonised parchmentised thread is composed. In some cases the original carbonisation of the material is performed in an electric furnace at a very high temperature and under pressure; under these conditions the organic material which is carbonised is converted into an exceedingly dense variety of carbon, called adamantine carbon, the surface of which has a smooth and polished appearance and a steel-like lustre without having been subjected to the above-described process of treating. However prepared, the carbon filament has then to be attached to the leading-in wires, which are sealed through the glass. These, as above stated, are made of platinum, and no really successful substitute has yet been found for this. The reason for this selection is as follows:—In order that the wires by which the current is conveyed to the carbon filament may be sealed into the glass airtight, and not shrink away from the glass when cold, it is necessary that the material of which the wires are composed shall have practically the same expansion as the glass, because otherwise the wires would crack out when the glass bulb was heated and cooled, and the air from outside would leak into the vacuum. It happens, by a fortunate coincidence, that the co-efficient of expansion of platinum is exactly the same as that of some varieties of glass, and platinum wires can therefore be sealed through the glass whilst hot, and will remain firmly attached to the glass when cold. The carbon filament is fastened to the ends of the platinum wires by means of a carbon cement, or a deposit of carbon made over the junction, and the carbon filament with its attached platinum wires is then sealed into a glass bulb. We are thus able to pass the necessary electric current through the filament, and yet preserve the carbon in a highly perfect vacuum.

In Fig. 18 are shown two lamps with carbon conductors of different forms. In some cases the carbon filament has a simple horse-shoe shape as originally adopted by Edison; whilst in other cases it is given a double twist, as in the Edison-Swan lamps, one or more curls being used. For some purposes it is necessary to coil the filament into a tight compact coil, in order that, when the lamp is used in the focus of a mirror, the light may be all concentrated in one place, and

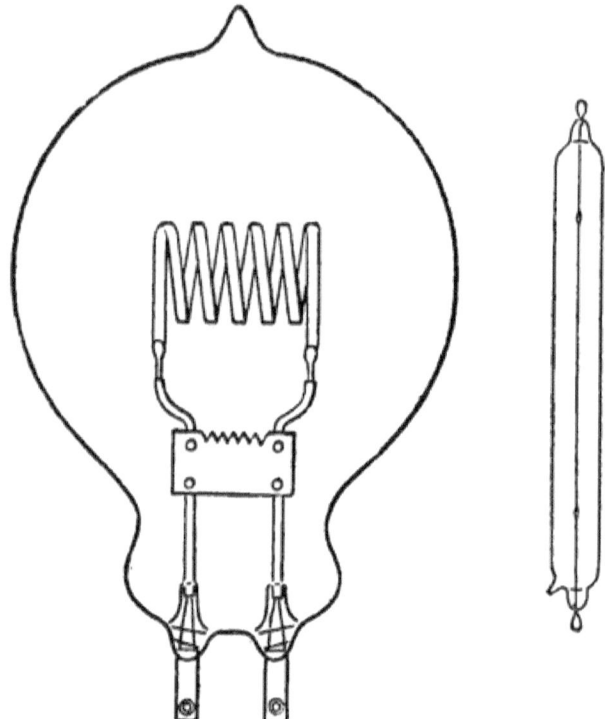

FIG. 18.—Glow Lamps with Zigzag and Straight Carbons

be gathered up by the mirror or lens. Such lamps are called *focus* lamps, and are employed in optical lanterns, in ships' side-lights, and wherever it is necessary to collect all the rays from the filament by optical means. These lamps are very useful in optical lanterns, and, now that so many houses are supplied with electric current, afford an easy means of showing lantern views in a drawing-room.* In cases in which high candle-power is required, it is obtained by putting two or more filaments in the same lamp bulb. In Fig. 19 is shown a representation of a multiple-filament lamp of 2,000 candle-

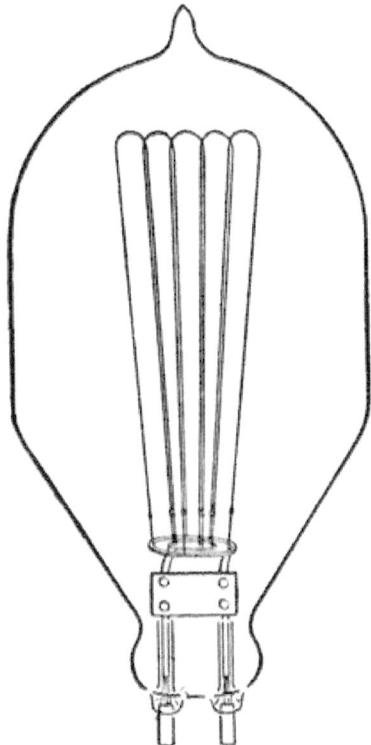

Fig. 19.—2,000 candle-power Glow Lamp with Multiple Filaments.

---

* Photographic amateurs who practise the art of making lantern slides from negatives, want an easy and ready mode of exhibiting them. If electric current is laid on to the house, the simplest method of doing this is to fit an ordinary optical lantern with a 100-volt focus lamp of 50 candle-power. This can be fitted on the tray which usually carries the oxyhydrogen jet. A screen of drawing paper six feet square makes an admirable surface for projection, and the views can be shown in a drawing-room without danger or smell.

power as constructed by the Edison-Swan Company. In these lamps a number of carbon conductors of the horse-shoe shape are arranged between two rings, so that the electric current can pass from one ring to the other through four, or five, or six, or more carbon conductors, each of which has one leg attached to one ring and the other leg to the other. Lamps of this high candle-power are made up to 3,000 candle-power or more. In such cases the glass bulbs become globes of glass nearly a foot in diameter, and have to be made of considerable thickness to withstand the enormous pressure of the air on the outside. The large lamp now shown to you in action has a globe having a surface of 280 square inches, and the air pressure on the exterior amounts to one and three-quarter tons! One limitation to the safe size of such lamps is this great external air pressure.

The final stage of manufacture of the lamp is the exhaustion of the lamp bulb. It was mentioned above that this was rendered essential, in the first place, by the fact that, if heated in the air to a high temperature, the carbon filament would soon be destroyed by combustion. Taking a non-exhausted lamp, I pass an electric current through the filament. You see that the lamp burns for a few seconds and then goes out. The oxygen of the air in the bulb has combined with the carbon of the filament in one or more places and destroyed it by producing combustion of the material. But if we take a filament of carbon not enclosed in a bulb and plunge it into a glass vessel full of a non-oxidising atmosphere, such as carbonic acid gas or coal gas, and heat it electrically, you will see that the carbon, though rendered incandescent, is not then rapidly destroyed. The carbon filament is enclosed in a vacuum, not merely to prevent its combustion, but there is another important reason for exhausting the bulb of all gases or air. Even if we suppose the bulb filled with a non-oxidising atmosphere of nitrogen or hydrogen, such a lamp would yet not be so perfect as a vacuum lamp, and for the following reason :—Modern physical research has provided arguments well nigh irresistible to prove that gases consist of molecules or little particles which are in rapid motion. This *kinetic theory* of the structure of gases, as it is termed, is supported by an immense body of physical evidence, and no less an authority than Lord Kelvin has called it one of the surest articles of the scientific creed. Certain lines of investigation have shown how to determine the *average velocity* of these gas particles. It has been shown, for instance, that in hydrogen gas at ordinary pressure and at the temperature of melting ice the molecules are moving with an average velocity of about 6,000 feet per second, or 69 miles per minute, and in oxygen with an average velocity of 18 miles per minute. These gas molecules in their rapid flight collide against one another in moving to and fro, and the average space over which the molecule flies before it has a collision against some other molecule is called its *mean free path*.

In one cubic foot of air or oxygen at the ordinary pressure, and at the temperature of melting ice, there are, in all probability, a number of gas molecules represented numerically by a billion times a billion, or by unity followed by twenty-four cyphers (10^{24}). The mean free path of the oxygen gas molecule is rather more than two-millionths of an inch, and the average number of collisions during the one second in which it darts over a distance in all of 1,500ft. is 7,600 millions. We are here dealing with minute portions of time and space, in which the unit is not a second and an inch, but a millionth part of each of these. The millionth of an inch is about the same fraction of an inch that one foot is of 200 miles; and the millionth part of a second is the same fraction of a second that one second is of about twelve days. All the air we breathe and feel consists, therefore, of these flying and colliding oxygen and nitrogen molecules. Each particle or molecule is always having its direction changed by its collisions against neighbours, just as a billiard ball has when it "cannons" against another. It is the constant bombardment of these molecules against the walls of a containing vessel which creates the elastic pressure of a gas, and when any solid body is surrounded by a gas an enormous number of collisions take place against that surface in every second of time.

If the surface is heated, the molecules which come up against it take heat from it in their myriads of contacts per second, and this removal of heat from the body by the gas molecules is called *convection*. If the carbon filament is enclosed in a glass bulb not exhausted of its air, a certain amount of energy is thus removed from the filament and given up to the glass, being conveyed by these molecular vehicles. The energy so abstracted from the filament is not in any way conveyed to the eye as light. Hence, for all illuminating purposes, it is lost, and a lamp so constructed with a non-vacuous bulb cannot in some respects be as efficient a device as one with a high vacuum, because of this additional loss of energy from the filament.

The process adopted for removing the air from the bulb is generally the employment of some form of mercury pump in combination with a mechanical air-pump. Time will not permit me to describe the many forms of this device which have been suggested. Briefly speaking, the principle upon which many of these mercury pumps act is as follows:—A vertical glass tube F (*see* Fig. 20) has a side tube S sealed into it, to which is attached the lamp to be exhausted. Drops of mercury are allowed to fall down the vertical tube, and as they fall down they push the air in the form of bubbles before them. The air is, therefore, gradually drawn out from the lamp bulb L, and the completion of the process of exhaustion is recognised

ELECTRIC GLOW LAMPS.

when the globules of mercury fall down the vertical tube in close contact with one another without carrying any air between them. Under these circumstances, if the bottom end of the vertical fall tube is placed in a vessel of mercury V, the mercury will gradually rise in the fall tube to the height

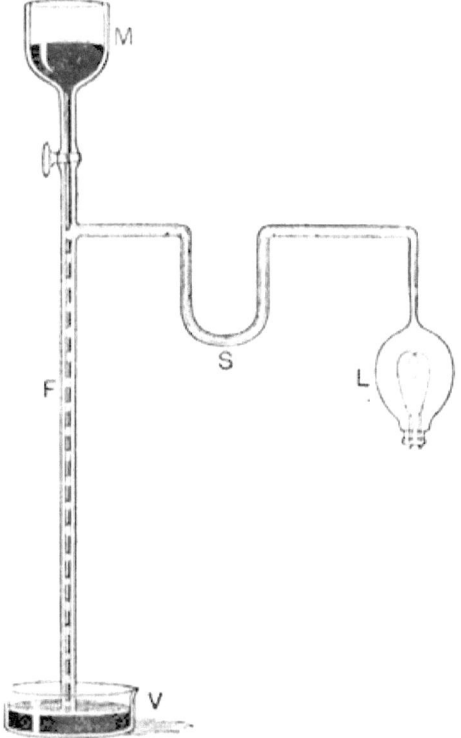

Fig. 20.—Diagram of arrangements of a Mercury Pump for Exhausting Lamps.

M Mercury Reservoir; F, Fall Tube; L, Lamp being Exhausted; S, Side Tube; V Vessel of Mercury.

of the mercury barometer—that is, to about 30in.—and then each succeeding drop of mercury which falls down on to the top of the column, does so with a sharp click, thus showing that the lamp bulb has been exhausted of its air to a very high degree. It is not difficult to abstract from such a glass lamp bulb all but one-millionth of the air originally contained in it. A highly exhausted space of this kind is commonly called a vacuum; nevertheless, when we realise that every cubic foot of air at ordinary pressure contains a billion times a billion molecules of air, it is easily seen that the reduction of the number of molecules which may be contained in a lamp bulb of about three inches in diameter before exhaustion, to one-millionth of their original value, still leaves included in the bulb many millions of millions of molecules. One important change that is, however, made in the physical state of the gas when so reduced to a very low pressure is that the *mean free path* of the molecules is enormously increased. By this reduction in pressure, the mean free path of the molecules in space, on an average, traversed before collision, is increased just in proportion as the pressure is diminished.

By the reduction of the pressure to one-millionth of its normal value by the abstraction of 999,999 millionths of the air, the mean free path of the air molecules is increased a million-fold—in other words, to something not far from two inches. In the interior, therefore, of a large glow lamp, which may have a volume, say, of ten cubic inches, there is good reason to believe that, when the bulb is exhausted to the highest point usually obtainable in practice, there still remain an immense number of molecules of air, but that the space is, comparatively speaking, so sparsely populated with molecules that each molecule in flying to and fro, may move, on an average, over a distance of an inch or two without collision against a neighbour. The rarefaction of the air must, therefore, decrease immensely the rate at which energy is taken from the filament by convection, because it decreases the number of molecular bombardments against the filament, and this is one object of producing the exhaustion. When the proper vacuum has been obtained in the lamp, the glass inlet tube is melted, and the lamp is sealed off from the pump. The lamp is finished by adding to it a brass collar with two sole-plates of brass, which are fastened on to the ends of the platinum wires which protrude through the glass. In this way an electric current can be sent through the filament, and yet the highly perfect vacuum in the glass bulb be indefinitely preserved.

Lamps so made may take numerous forms. In Fig. 21 is shown a form of lamp called a candle-lamp, employed

FIG. 21.—Electric Candle Lamp, of which the Glass Bulb is shaped like a Candle Flame.

for decorative and other purposes. Lamps may be made so small that they can be employed for theatrical, decorative, microscopic, dental, and surgical purposes, in which case they are called *micro-lamps* ; they may be made so large, and have so many carbon conductors in them, that they can yield a candle-power of two to three thousand, and the glass bulbs may, in both cases, be given any requisite form and colour.

Having thus briefly described the process of the construction of a glow lamp, we have now to study a little more closely the laws which govern its operation. From what has been already stated in the First Lecture, it will be seen that, in order to possess a full knowledge of the behaviour of the lamp as an energy-translating device, it is necessary to know four things about it. First, we must know the current, in amperes, which is taken by the lamp when a certain electric pressure is produced between the ends of the carbon filament. Generally speaking, lamps are denominated by the pressure at which they must be worked to give satisfactory results as to duration. This pressure measured in volts is called the *marked volts* or *maker's volts*, and lamps may be distinguished as 100-volt lamps, 50-volt lamps, 10-volt lamps, &c. The current taken by a lamp can be measured by many kinds of instruments, which are called *ampere-meters* or ammeters. The pressure between the terminals of the lamp can be measured by suitable *voltmeters*. Approximately speaking, a lamp intended to work at 100 volts and to give 16 candles light has a carbon filament about five inches long and about one hundredth of an inch in diameter, and when worked at a pressure of 100 volts it takes from $\frac{3}{5}$ to $\frac{2}{3}$ of an ampere. From explanations previously given it will be remembered that if these two numbers—namely, the number representing the volts and the number representing the amperes—are multiplied together, we have a number representing the *electrical power*, estimated in watts, which is being taken up in the lamp, and such a 16-candle-power 100-volt lamp generally takes from 50 to 60 watts. We have already described the methods by which the candle-power can be measured. If we divide the number representing the electrical power in watts taken by the lamp by the number representing the candle-power of the lamp, the quotient gives us the number representing the *watts per candle-power*. In most modern lamps this last number varies from $2\frac{1}{2}$ to $3\frac{1}{2}$, when the lamp is being used at the marked volts. There are, therefore, four quantities in

connection with every lamp which it is important to know.
First, the *marked volts* or voltage of the lamps; second,
the *ampere current* taken by the lamp at the marked volts;
third, the *candle-power* of the lamp; and, fourth, the *watts
per candle-power*. Very briefly we may describe the process
by which these four quantities are determined.

We have already described the instrument called an electrostatic voltmeter, which is used for measuring electric pressure.

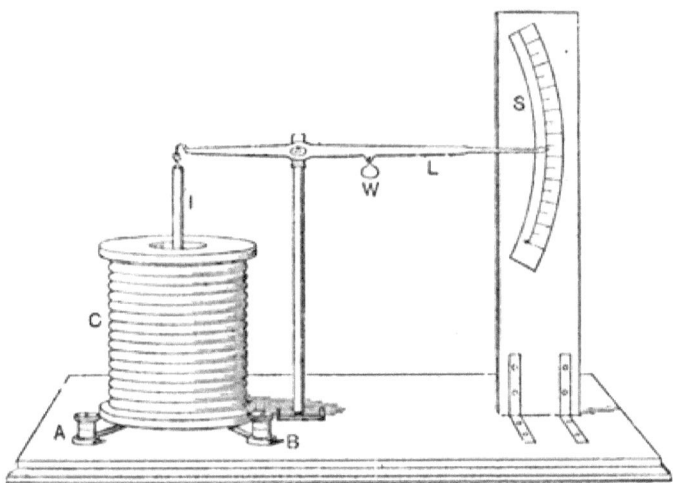

FIG. 22.—Simple Form of Ampere-Meter for Measuring Lamp Currents.

One simple form of instrument, called an ammeter, which
may be used for measuring the current in amperes taken
by a lamp, is made as follows :—A coil of wire **C** (*see* Fig. 22)
is wound on a bobbin, and through this coil of wire the
current to the lamp is passed, This current creates a magnetic field round the wire, as described in our First Lecture.
If a piece of iron wire I is suspended on the end of a
delicately-balanced lever, so that the iron is held freely just
at the entrance of the bobbin, the passage of an electric
current through the bobbin will attract the iron into the coil.

This movement of the iron is resisted by a weight, W, properly placed on the balance arm, or by the weight of an indicating needle. Such an instrument may be graduated to show on a divided scale the strength of the current in amperes which is passing through the coil.

It is important that, in thus making a determination of the constants of a lamp—viz., the current passing through the lamp, the pressure difference at the terminals of the lamp, and the candle-power of the lamp by comparison with a suitable standard of light—these quantities should all be measured at the same instant. By varying the current through the lamp, as we can do by using pressures of different values, we can obtain a series of observations in which the candle-power, current, and volts on the terminals of the lamp are all measured simultaneously for different values of the pressure on the lamp terminals, and these values can be set out in a series of curves which are called the characteristic curves of the lamp. In Fig. 23 is shown a curve illustrating the manner in which the candle-power of a 16-c.p. lamp varies with the current. It will be seen that a very small increase in the current is accompanied by a large increase in the light of the lamp, and it can be shown that the candle-power varies very nearly as the sixth power of the current. That is to say, if a lamp has an illuminating power equal to one candle when a current of one ampere is passing through it, then it would give $2 \times 2 \times 2 \times 2 \times 2 \times 2 = 64$ candle-power if two amperes were passed through it. No carbon filament lamp could endure without destruction an increase in current to the extent of 100 per cent., but within narrower limits the above law holds good, viz., that the candle-power varies approximately as the fifth or sixth power of the current. In Fig. 24 is shown another curve, illustrating the manner in which the candle-power varies with the volts, and in Figs. 25 and 26 curves showing the manner in which the candle-power varies with the watts and with the watts per

ELECTRIC GLOW LAMPS. 71

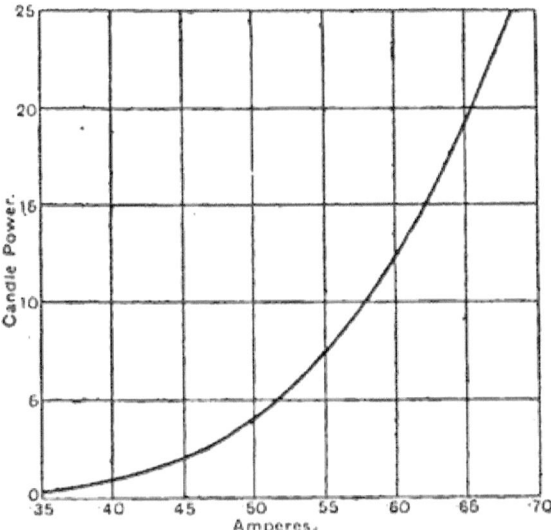

FIG 23.—Curve showing Variation of Candle-power with Current in an Electric Glow Lamp.

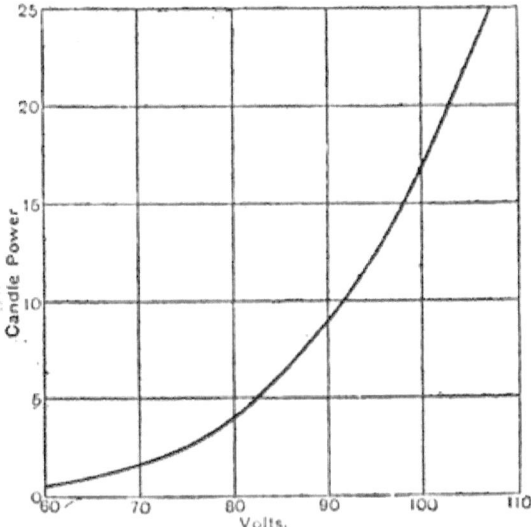

FIG. 24.—Curve showing Variation of Candle-power with Voltage in an Electric Glow Lamp.

candle-power. From the curve in Fig. 25 it will be seen that the candle-power of the lamp varies very nearly as the cube of the total power in watts supplied to it. In other words if we supply the lamp with twice the electrical power, the candle-power is increased eight times. If we supply it with three times the electrical power, it is increased twenty-seven times. These curves will show, amongst other things, the importance of preserving the pressure at the terminals of a carbon filament electric lamp constant when used as an

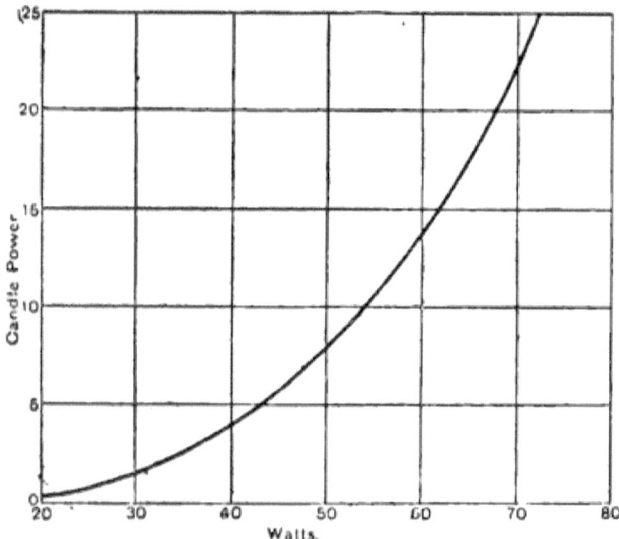

Fig. 25.—Curve showing Variation of Candle-power with Wattage in a Glow Lamp.

illuminating agent. Any variation of the pressure, however small, produces a very serious variation in the candle-power of the lamp. It is convenient to bear in mind the following figures for the 16-c.p. lamp as generally made. Such a lamp takes, when worked at a pressure of 100 volts, 0·6 or $\frac{3}{5}$ of an ampere, absorbs 60 watts of power, and therefore in $16\frac{2}{3}$ hours consumes an amount of energy which is $16\frac{2}{3}$ times $60 = 1,000$

watt-hours. This amount of energy is called one Board of Trade Unit, and is generally sold by the Electric Supply Companies in England at prices varying from 3d. to 8d.

As soon as carbon filament electric lamps came into general use it was found that lamps are like human beings —they have a certain life or duration, and moreover most of them undergo a process of deterioration in light-giving

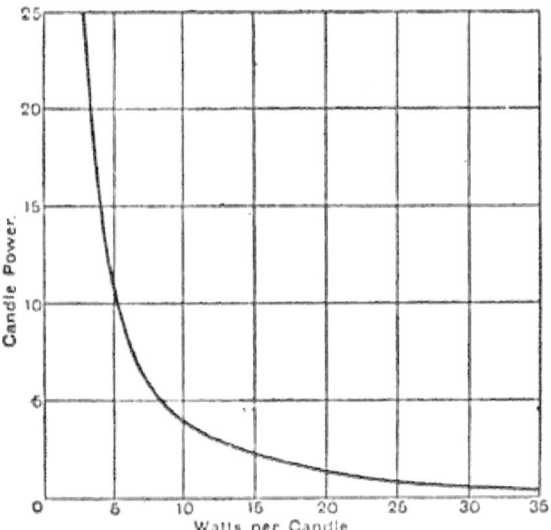

Fig. 26.—Curve showing the Variation of Candle-power with Watts per Candle-power in an Electric Glow Lamp.

power. After a certain time the filament of the lamp is destroyed, and the lamp ceases to work. The duration of any lamp when worked under certain conditions as to pressure and supply is called its *life*, and from a large number of observations of similarly-constructed lamps when worked in a particular manner we can deduce the *average life* of a lamp. We cannot predict from observations of one or two lamps what the average duration will be, any more than we can tell from observations on one or two human lives what is the average duration of human life in any place. The life of any particular individual lamp may be only a few hours, or it may be many thousands of hours. We shall presently point out that this average duration is only one factor in determining the real value of the lamp as an energy-translating device. Apart from accidental circumstances, the factors that determine the life of the lamp are the temperature at which the carbon filament is kept, and the method of its manufacture. Assuming, however, that an incandescent lamp is supplied with electric current at a constant electric pressure: it is always found that two marked changes occur in the lamp in the process of time; first, the lamp begins to blacken by a deposit of carbon which is made upon the interior of the bulb; and, second, the carbon filament undergoes a change by which its electrical resistance becomes increased and its surface altered, so that, whatever its initial condition, its surface finally assumes a much darker and more sooty appearance than when first manufactured.

From these physical changes it follows that the candle-power of the lamp is diminished, first by an obstruction of light by the black deposit of carbon on the inner surface of the glass, and, second, owing to the increase in the resistance of the filament, the lamp taking less current and therefore furnishing less light. The deposit of carbon on the interior of the bulb is produced partly, or perhaps generally, by a process which is an ordinary evaporation of the carbon, but it may be produced also by volatilisation of condensed hydrocarbons from the interior of the conductor. In addition, however, to this, there is a process of projection of carbon molecules from the filament in straight lines which is not of a kind generally included in the term volatilisation. Many blackened carbon lamps show lines of no deposit on the glass, which

ELECTRIC GLOW LAMPS. 75

have been called *molecular shadows*. It is not unusual to find lamps the interior surface of the glass of which has become very black by a deposit of carbon. On examining the lamp it will be found that on one side of the glass, lying in the plane of the loop, there is a line of clear glass on which no deposit has been made. This is very often shown well in lamps in which the filament has been worn away at a point about halfway down one leg of the loop (*see* Fig. 27); and on examining such a lamp it will be found that there is a well-marked line of no deposit upon the glass, just in the

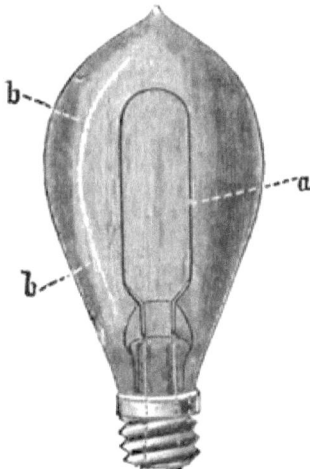

FIG. 27.—Glow Lamp with the glass bulb blackened by deposit of carbon, molecular scattering having taken place from the point marked *a* on the filament, and a shadow or line of no deposit thus produced at *b* on the glass receiver.

plane of the loop, and on the side farthest removed from the point of rupture. Blackened lamps of this kind, therefore, indicate clearly that carbon molecules have been shot off in straight lines from the place where the carbon has subsequently been ruptured. If in a carbon filament there happens to be a weak spot or place of high resistance, at that point the current will generate heat at a greater rate than at other places, and the temperature of that spot will rise above the temperature of the rest of the filament. Such points of greater temperature can often be seen on carefully examining an old or bad filament. From these high-temperature spots, carbon molecules must be projected off by an action which may be partly electrical, and are deposited upon the glass on all sides. But it is clear that the unbroken leg would shield part of the glass from this carbon molecule bombardment, and a *shadow* of one-half of the loop would therefore be formed upon the glass.

This shielding action may be illustrated by a simple experiment. Here is a ∩-shaped rod (Fig. 28). This shall represent

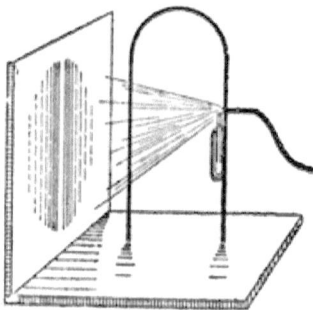

Fig. 28.—"Spray Shadow" of a rod thrown on Cardboard Screen to illustrate formation of molecular shadow in Glow Lamps.

the carbon conductor in the lamp, and this sheet of cardboard placed behind it the side of the glass bulb. I have affixed a little spray-producer to one side of the loop, and from that point blow out a spray of inky water. Consider the ink spray to represent the carbon atoms shot off from the overheated spot. We see that the cardboard is bespattered on all points except along one line where it is sheltered by the opposite side of the loop. We have thus produced a "spray-shadow" on the board. The existence of these molecular shadows in incandescent lamps leads us, therefore, to recognise that the carbon atoms must be shot off in straight lines, or else, obviously, no such sharp shadow could thus be formed. It has been asserted by some writers that the whole of the black coating on the interior of the carbon glow lamps is produced solely by a process of evaporation of carbon; but the frequent existence of these shadows on blackened lamps show that under some conditions the projection of carbon from the filament is not merely a general and irregular volatilisation, but a copious projection of carbon molecules, which move out in perfectly straight lines from one part of the filament.*

It is clear, therefore, from all the foregoing considerations connected with the processes which are going on in the interior of the lamp and the resulting ageing of the filament, and from the changes that take place in the candle-power with change of current, that the following conditions must be observed in order to secure the best results in the employment of incandescent lamps for illuminating purposes. In the first place, the electric pressure which is supplied to the consumer must be exceedingly constant. Where a supply is drawn from the mains of a public electric lighting company the consumer will, usually, have no control over the pressure, at any rate in the case of continuous currents. The public supply companies of Great Britain are bound by their Electric

---

* The author possesses a large collection of lamps showing these molecular shadows. In the case of the old Edison lamp, the bamboo filament was attached to the platinum leading-in wires by a deposit of copper made over the clamp. Under these conditions, if one clamp became loose, or made a bad contact, a scattering of copper molecules took place from that clamp, producing a green transparent deposit of copper over the interior of the bulb, which deposit generally showed a well-marked shadow of one-half of the carbon loop upon the glass. Shadows have also been produced by the author by the volatilisation of aluminium plates in lamps, the volatilisation having been created by passing a sudden strong current through an aluminium wire or plate included in the lamp bulb. This aluminium deposit is of a fine blue colour.

Lighting Orders to furnish current at a certain constant pressure, which is generally 100 or 110 volts, and conditions are laid down by the British Board of Trade that the pressure shall not vary more than four volts above or four volts below this specified pressure. These are very wide limits of variation, and very few of the supply companies avail themselves of them. The consumer who takes current from a public

FIG. 29.—Holden's Self-recording Hot-Wire Voltmeter.

supply company should always take pains to ascertain for himself whether the pressure *is* constant. He can do this by the employment of a voltmeter, such as Lord Kelvin's electrostatic voltmeter, described in the previous Lecture, Fig. 8, or he can employ a self-registering voltmeter, of which there are many in use, which will record upon a drum covered with paper a curve indicating the variation of pressure during the day and night. Two of these recording voltmeters which are now much in use are those devised by Prof. G. Mengarini and Captain Holden. In the latter instrument (*see* Fig. 29), a very fine wire is connected across between the two points between which pressure has to be determined. A small current flows through this wire and heats it, causing it to expand. The expansion of this wire is then made to move a lever, which, by means of a writing pen, records upon a drum driven by clockwork a curve, and any variation in the pressure or voltage is shown by the form of the line so drawn. The other instrument, due to Prof. Mengarini, is of a different type. In this apparatus a current taken from the two points between which the pressure has to be measured is passed through two coils of wire—one a fixed coil, and the other a coil suspended by two steel wires. The fixed and movable coils are so arranged (*see* Fig. 30) that when a current passes through the two it causes the movable coil to twist round, or tend to twist round, into a new position. This action of the current is resisted by the suspended wires. A writing pen is attached to the movable coil and draws a curve upon a revolving drum, as above described, and this curve records any variation in the voltage. The chart so drawn can then be examined at the end of every day, and a record is thus kept of the variation of pressure during the twenty-four hours. If the consumer finds that the light of his incandescent lamps varies a great deal, or if the life of the lamps is very irregular, it is well to have such a curve of pressure automatically drawn by a self-recording voltmeter for three or four days, and in this way to detect the limits within which the electric pressure varies. Consumers are very apt to lay the blame immediately upon the lamps for any variation in light or brevity in life, but as the electric pressure is partner with the lamp in producing the result, it is evident that the sins of the supply companies in giving a variable pressure ought not to be laid upon the lamps. On the other hand, if the recording voltmeter shows a great uniformity in electric pressure, and if still the life of the lamps is not satisfactory, then either the lamps are in fault, or else the consumer is not using lamps of the proper voltage for his circuits. Before ordering lamps, it is well to ascertain also in this way what is the actual pressure

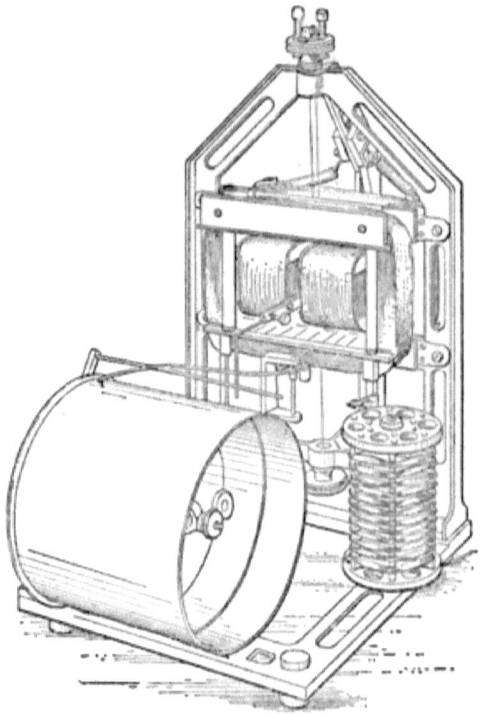

Fig. 30.—Mengarini's Self-recording Voltmeter.*

on the circuits, and then to order lamps of the proper marked volts for the particular pressure. Incandescent lamps which have been made and marked by the maker to work at 100 volts are not intended to be worked at 102 or 104, and, if they are, an abbreviation in the average life must certainly

---

* Reprinted by permission from the *Society of Arts Journal*.

occur, or at any rate a more rapid blackening of the globe and diminution of the candle-power than would be the case if the consumer employed lamps of the proper voltage. If the supply is very irregular in pressure, it is never possible to obtain the same good results in the duration of the lamp and uniformity in brilliancy which can be obtained if the supply is maintained constant in pressure, at least within the limits of one or two volts.

Let us at this point note that, apart from the question of duration or cost, the physical value of a glow lamp is

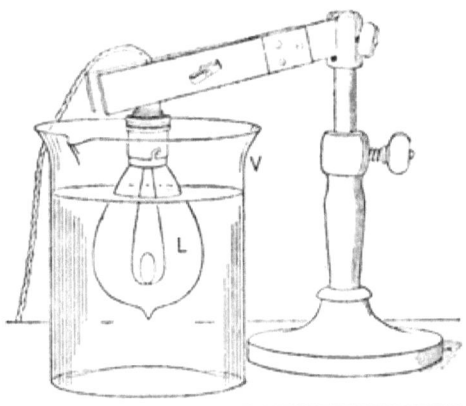

FIG. 31.—Glow Lamp Burning under Water.

measured by its utility as an energy-transforming device. An electric glow lamp is a machine for transforming energy from one form, viz., the energy of an electric current, into another form, viz., eye-affecting radiation. As explained in Lecture I., it transforms the electric energy into radiant energy, or energy of ether waves, only a fraction of which can impress the eye as light. This fraction varies from 3 to 7 per cent. This numerical expression for the "luminous efficiency" can be determined by suitable experiments in which the relative amounts of luminous and dark heat sent out by a glow lamp are separately measured. If a glow lamp L (*see* Fig. 31) is immersed in water in a transparent vessel V the water absorbs or stops nearly all the dark or non-luminous radiation, but permits a large fraction of the luminous radiation or light to pass out. By photometering the lamp both when in and when out of the water the fraction of the luminous radiation which is absorbed can be determined. By measuring the rise of temperature of the water which is created by the absorption of the radiation, both when the luminous radiation is allowed to pass out and when it is wholly stopped by blackening the lamp, we can determine the ratio between the amount of the radiation which affects the eye as light and that which cannot affect it at all. A well-devised series of experiments of this kind have been carried out by Mr. E. Merritt, at Cornell University, U.S.A. The result has been to show that, on an average, about 5 per cent. of the energy supplied to the glow lamp as ordinarily used is utilized to make light, and the other 95 per cent. is non-effective as far as the eye is concerned. Hence, if a glow lamp is supplied with 60 watts and gives 16 candles light, only 3 watts are really transformed into light, and the mechanical value of the light so produced is rather less than a quarter of a watt per candle.

This luminous efficiency varies with the temperature of the filament. At a low red heat it is not more than 1 per cent., but increases to about 5, 6, or 7 per cent. at the normal working temperature. Thus, for an Edison 16-c.p. lamp Mr. Merritt found the following figures when the lamp was worked at various pressures so as to give different candle-power.

It will be seen that the luminous efficiency increases as the "watts per candle-power" decreases—that is, as the temperature of the filament increases.

The "luminous efficiency" must be carefully distinguished from that which is sometimes called "the efficiency" of the lamp, and which is the reciprocal of the watts per candle-power or the candle-power yielded per horse-power or per watt expended in the filament.

Observations on the Luminous Efficiency of a 16-c.p. Edison Glow Lamp.

Working volts.	Candle-power.	Power absorbed in watts.	Watts per candle-power.	Luminous radiation in watts.	Luminous efficiency as a percentage.
74·2	0·9	34·6	38·0	0·18	0·5 per cent.
91·6	4·8	56·2	12·0	0·68	1·2 ,,
97·3	7·3	64·6	9·0	1·13	1·7 ,,
100·3	8·9	69·3	7·8	1·62	2·3 ,,
107·6	14·6	81·6	5·6	2·97	3·6 ,,
109·3	16·3	84·4	5·2	4·57	5·4 ,,
124·1	38·2	115·4	3·0	7·46	6·5 ,,

It will be convenient at this stage of our discussion to enter a little more fully into the changes that go on in an incandescent electric lamp during its life. We have pointed out above the causes which are at work to change the surface and ultimately to destroy the carbon filament. These causes commence to act from the very moment when the lamp begins to be used. If a carbon filament electric glow lamp is placed upon a circuit of perfectly constant pressure, and if the constants of the lamp—namely, the candle-power, the current, the resistance of the filament, the power taken up in watts, and the watts per candle-power—are measured at intervals, say of every 100 hours, it will be found that the changes which go on in the lamp are somewhat as follows:—During the life of the lamp there is a progressive rise in resistance, a progressive diminution in candle-power, and a progressive increase in watts per candle-power. An illustration of these changes is diagrammatically shown for certain forms of ordinary vacuum lamps by the curves given in Fig. 32. These curves are deduced,* not from observations on one lamp alone, but from the average of a large number. If, instead of working the lamp at its marked volts or normal pressure, we raise the voltage, say, 10 per cent. above the normal, changes of the same character take place, only they are very much accelerated, and the result for a collection of lamps worked at 10 per cent. above the normal pressure is given in Fig. 33.

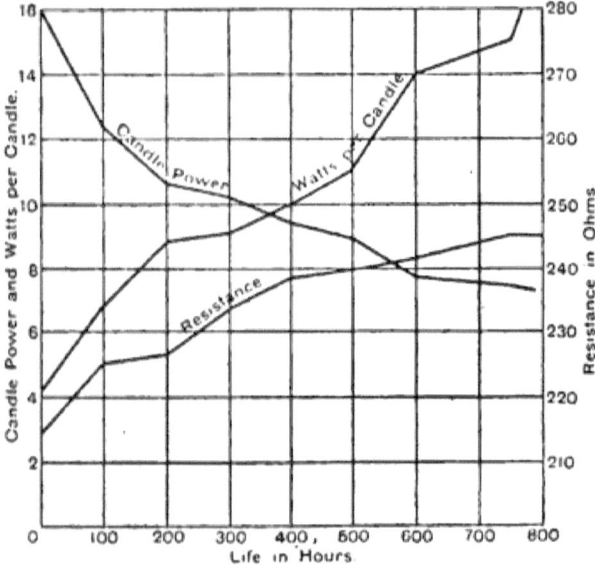

Fig. 32.—Curves showing the increasing resistance and diminishing candle-power of a Glow Lamp during its life.

On the other hand, if we desire to keep the candle-power always constant, we have to continually raise the pressure, and at the same time we are increasing the watts per candle and the resistance of the lamp. These progressive changes are illustrated in the two curves in Figs. 34 and 35, in the

---

* These curves are borrowed from a very instructive Paper by Prof. Nichols, of Cornell University, U.S.A., on "The Artificial Light of the Future."

first of which a 16-c.p. lamp was worked at continually-increasing pressures sufficient to keep the candle-power always constant, and in the second of which the candle-power was raised to four times its normal value and kept constant.

The value of a lamp to a user is not to be measured merely by the watts per candle-power it takes at the commencement

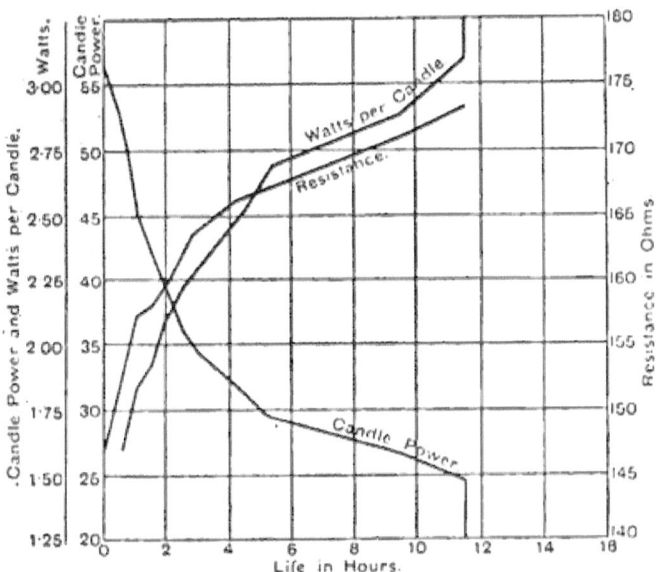

Fig. 33.— Curves showing the rapid decrease in candle-power and increase in resistance in a Glow Lamp worked at 10 per cent. above its normal pressure.

of its life, but also, to a large degree, by the extent to which this consumption of energy and production of light remains constant during the life of the lamp. This is best expressed by the percentage of the variation of the candle-power during the life of the lamp when taken at equal intervals of time. A 16-c.p. glow lamp, taking say 60 watts at the commencement of its life, but which after two or three hundred hours in use is giving no more than 8 candles light, although taking the same power, is costing the consumer twice as much per candle at the end of this time as at the beginning. Some writers have, therefore, suggested that at the end of a certain period, when the candle-power has suffered a definite diminution in value, say, of more than 30 per cent., the lamp should be "smashed," and a new lamp put in its place. This, however, is not always advisable or necessary.

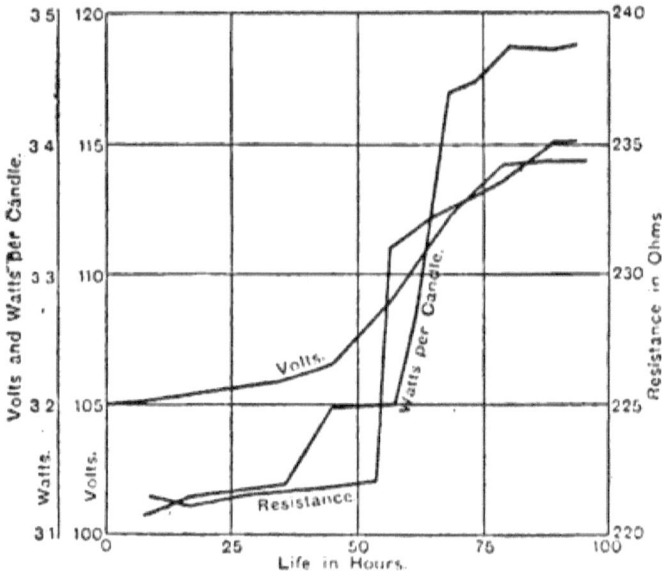

Fig. 34.—Curves showing the increase in Voltage required to keep the candle-power of a Glow Lamp constant.

If the lamp in the position in which it is being used is giving light enough for the purpose for which it is required, then, although its candle-power may be diminished, if the total power taken by the lamp is diminished also, the total sum which is being paid by the consumer for the energy supplied to that lamp is on the whole diminished, and therefore the lamp is costing the consumer less *on the whole* to maintain in action than a new lamp would do if it replaced the old one, although at the same time the cost of the lamp per candle-light may be considerably increased. Generally speaking, users of electric glow lamps have an idea that the lamp which lasts the longest is necessarily the best. From what has been stated above, it will be seen that it is not merely long life, but constancy of candle-power, combined with high luminous efficiency and low cost, which are really the factors deter-

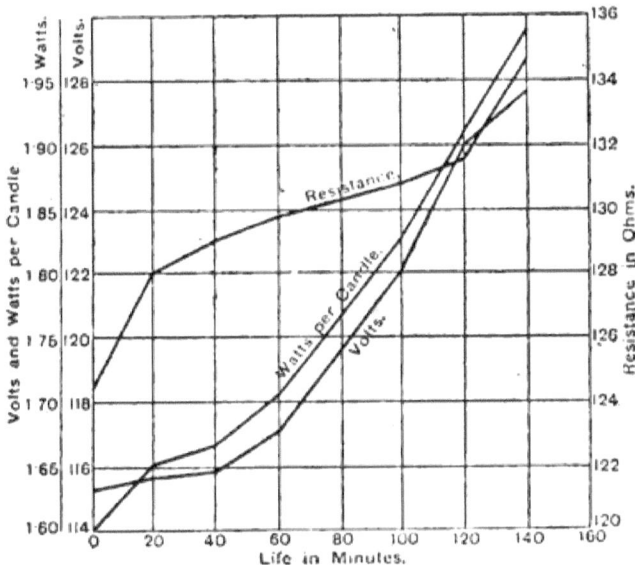

FIG. 35.—Curves showing the increasing Voltage necessary to keep the candle-power of a Glow Lamp constant and equal to four times its normal amount.

mining the value of the lamp to the consumer. The duration of the lamp is intimately connected, apart from accidents, with the temperature at which the carbon filament is maintained, and this is determined by the watts per candle-power being expended in the lamp. Makers of glow lamps sometimes make statements which lead the uninitiated to believe that there is a certain "efficiency" or peculiarly low value of the watts per candle-power belonging especially to a certain make of lamp. It should be, however, clearly understood that any glow lamp may be worked at values of the watts per candle-power varying over very wide limits, but experience has shown that carbon filaments, as at present made, cannot be worked at temperatures higher than those corresponding to about 2 to $2\frac{1}{2}$ watts per candle-power without very rapidly being destroyed.

A number of tests have been made by Messrs. Siemens and Halske on lamps of various types, and in which groups of these lamps were worked at different watts per candle-power from 2 to 4 and upwards. The lamps were measured at intervals of 100 hours, and the diminution in the observed candle-power expressed as a percentage of the original candle-power was laid down in the form of curves. The results of these tests seemed to show that, on the whole, lamps worked at from 3 to $3\frac{1}{2}$ watts per candle-power had a greater average constancy in candle-power than those worked at a lower number of watts per candle-power. That is to say, if we take a 16-c.p. glow lamp and work its filament at such a temperature that it is taking on the whole 32 watts, or 2 watts per candle-power, it may take less power to begin with, but it will not preserve nearly the same constancy of candle-power as if it had been worked at such a temperature that it absorbed from 48 to 56 watts at the commencement of its career; in other words, absorbed 3 to $3\frac{1}{2}$ watts per candle-power. No hard and fast rule, however, can be laid down about these matters. No deductions can be made which have any great value from tests made upon a small number of lamps. It has already been stated that the lives of electric lamps are in this respect like human lives, and just as the insurance companies have to draw their deductions and to make their calculations for life insurance on the results of figures obtained from a very large number of human lives, so inferences made from the results of statistical tests carried out on incandescent lamps are only of use if they are made on a very large number of lamps. Broadly speaking, however, the duration of an incandescent lamp is dependent upon the electric pressure at which the lamp is worked, and if a lamp is marked to be used on a 100-volt circuit, then if it is used on a circuit at a pressure of 105 volts, its life would be greatly abbreviated, probably to one quarter or one fifth of the time which it would last if worked at the normal pressure. On the other hand, if it is worked at a pressure of 95 volts, its duration, accidents apart, will be very greatly increased. It is useful, therefore, to have, under these circumstances, a general knowledge of what would be the variation in candle-power if the same lamp is employed at different voltages. The tables on the next page give the candle-power yielded by two particular electric glow lamps when worked at varying pressures, the first table being for a 100-volt 16-c.p. lamp, and the second table being for a 100-volt 8-c.p. lamp. A very cursory examination of these tables will show that the light given by the lamp varies enormously with the pressure, and that a very small diminution of pressure suffices to deprive the user of a very large fraction of the candle-power for which the lamp is constructed.

It may not be inappropriate at this point to call the attention of users of incandescent lamps to the fact that price is not the only factor to be considered in connection with the choice of a glow lamp. Consideration should also be given to the rate at which the candle-power decays when constant pressure is kept on the terminals of the lamp, as well as to the initial efficiency, and the *useful* life of the lamp. The useful life of the lamp, so far as most consumers are concerned, must be considered to be limited to that period of the total lamp life in which its candle-power is not diminished by

Tables *showing the Candle-power of certain nominal 8 and 16 Candle-power Incandescent Lamps when worked at various pressures in Volts.*

A 16 Candle-power Lamp.		An 8 Candle-power Lamp.	
Worked at	Yielded a candle-power of	Worked at	Yielded a candle-power of
105 volts.	22·8 c p.	102 volts.	7·9 c.p.
100 ,,	16·7 ,,	100 ,,	6·9 ,,
95 ,,	12·2 ,,	98 ,,	6·1 ,,
90 ,,	8·7 ,,	96 ,,	5·35 ,,
85 ,,	5·9 ,,	94 ,,	4·7 ,,
80 ,,	4·0 ,,	92 ,,	4·2 ,,
75 ,,	2·5 ,,	90 ,,	3·65 ,,
70 ,,	1·5 ,,	88 ,,	3·2 ,,
65 ,,	0·8 ,,	86 ,,	2·8 ,,
60 ,,	0·6 ,,	84 ,,	2·35 ,,
		82 ,,	2·2 ,,
		80 ,,	1·7 ,,
		78 ,,	1·4 ,,
		76 ,,	1·15 ,,
		74 ,,	0·9 ,,

more than 30 to 40 per cent. of its original candle-power, although this is not by any means a definite limit to the utility of the lamp. The user has to determine what is the total cost to him, during that useful life, of the energy supplied to the lamp, and what is the total candle-power hours given by the lamp. Suppose, for example, that the lamp is a 100-volt 16-c.p. lamp, and that it takes originally 50 watts to bring it to this candle-power, when on a 100-volt circuit. Then let us suppose, in the course of 600 hours' burning, that it is reduced to 8 c.p., and that it then takes 40 watts at the same pressure. If the rate of decay is tolerably uniform, the total energy taken by the lamp in watt-hours is the product of $45 = \frac{1}{2}(50+40)$, and 600, or 27,000 watt-hours; which is 27 British Board of Trade units (frequently written B.T.U.). The average candle-power hours given by the lamp is $\frac{1}{2}(16+8) = 12$, which multiplied by 600 gives 7,200 candle-power hours. If the Board of Trade unit of electric energy costs 6d., then the total cost of the energy taken up is 13s. 6d., for which the consumer gets 7,200 candle-hours of light at a mean candle-power of 12. If the lamp originally cost 1s. 6d., the total cost of the 7,200 candle-hours, including lamp, is 15s., or the consumer gets 40 candle-hours for a penny, given at a mean candle-power of 12. In other words, he is paying 0·3 of a penny per hour for his lamp and light. A penny includes, therefore, the cost of lamp and power for about $3\frac{1}{3}$ hours. Suppose, however, the user is offered a lamp at the price of 1s., but that extended experience shows that a 16-c.p. lamp of this kind falls to 8 candle-power in 300 hours' use. It is clear that this last lamp, if it takes the same power at each candle-power as the other, will transform in this time $45 \times 300 = 13,500$ watt-hours, or $13\frac{1}{2}$ Board of Trade units, and the total cost of lamp at 1s. and power at 6d. per unit to the user will be 7s. 9d. For this he gets $12 \times 300 = 3,600$ candle-hours, or 39 candle-hours for a penny. Hence, the cheaper lamp has not really much benefited him.

Therefore, the question which is so commonly asked, viz., Is electric light cheaper or dearer than gas? cannot be answered by a simple "Yes" or "No." The factors which decide the relative cost, so far as the electric glow lamp is concerned, are as follows:—1st. The price paid per Board of Trade unit of electric energy. 2nd. The price of the lamp. 3rd. The average useful life of the lamp. This last will depend not only upon the lamp itself, but also upon the constancy of the electric pressure of the supply. 4th. The manner in which the "wiring" of the house or building has been done; a factor which greatly determines the degree to which the consumer can effect economy by not using more lamps than is actually necessary. 5th. The intelligence shown in placing the lamps so as to make each do the utmost illuminating duty, and the

92   ELECTRIC LAMPS AND ELECTRIC LIGHTING.

care with which the control is effected by switching off lamps not actually needed. We may add also a 6th. The employment of a by-no-means negligible precaution in securing the best results—the keeping of the outside of the lamps clean and free from dust and dirt by having them well washed at intervals. A dusty lamp gives much less light than a clean one.

To assist the reader in making certain essential comparisons, a table is given below which shows the total cost in power and lamp renewals for working for 200 hours a 16-c.p. 100-volt glow lamp taking on an average 50 watts. The costs are given for various prices of the Board of Trade unit from 3d. to 8d., and for lamps costing 1s. and 1s. 6d., and for useful lives of from 200 to 1,000 hours. Let it be borne in mind that the "useful life" of a lamp is not a sharply marked period, but is the approximate average time during which its candle-power has not deteriorated by more than say 40 per cent. of its original value.

Table *showing the cost of 200 hours' use of a 100-volt 16-c.p. Glow Lamp, including the proportionate cost of Lamp Renewals and cost of Power. The Lamp is assumed to take on an average 50 watts, and to cost either 1s. or 1s. 6d.*

Cost of Energy. Board of Trade Unit.	3d.		4d.		5d.		6d.		7d.		8d.	
Price of Lamp.	1/-	1/6	1/-	1/6	1/-	1/6	1/-	1/6	1/-	1/6	1/-	1/6
Useful Life of Lamp in hours.	Cost of 200 hours' use of Lamp.											
200 hours.	3/6	4/-	4/4	4/10	5/2	5/8	6/-	6/6	6/10	7/4	7/8	8/2
400 "	3/-	3/3	3/10	4/1	4/8	4/11	5/6	5/9	6/4	6/7	7/2	7/5
600 "	2/10	3/-	3/8	3/10	4/6	4/8	5/4	5/6	6/2	6/4	7/-	7/4
800 "	2/9	2/10	3/7	3/8	4/5	4/6	5/3	5/4	6/1	6/2	6/11	7/-
1,000 "	2/8	2/9	3/6	3/7	4/4	4/5	5/2	5/3	6/-	6/1	6/10	6/11

The foregoing table must be read as follows:—Suppose the user gets current at 6d. per unit, and buys lamps at 1s. 6d., and let the lamp he buys be assumed to have an average useful life of 800 hours. On looking along the column in line with 800 and under the heading 6d. and 1s. 6d., will be seen 5s. 4d. This is the cost of the 16-c.p. lamp for 200 hours of burning, including proportion of lamp cost and power. Hence, in this case the 800 hours' burning cost 21s. 4d. on the whole, and for this the user gets an average candle-power of about 13 candles for 800 hours, or $13 \times 800 = 10,400$ candle-hours. It is obvious that no hard and fast comparison with the cost of equivalent candle-hours given by gas light is possible, unless a careful statement is made of the manner in which the two lights are used. One great advantage which electric illumination possesses, from a domestic and economical point of view, is the ease with which it can be turned off and on. A switch placed near the door enables a person leaving a room to turn out all the electric lights without calling for the smallest trouble in relighting them, which cannot be done in the case of gas. In effecting a comparison of cost, we have to take into consideration the time of wastage and non-useful combustion in the consumption of gas; and this, apart altogether from the immense superiority of the illuminant which fouls no air and destroys no decorations. Actual extended experience shows* that, with care and good management, electric energy at 8d. per unit can be made to give as much illumination by means of incandescent lamps as gas at 3s. 2d. per thousand cubic feet used with ordinary gas burners. At this rate, one Board of Trade unit of electric energy can be made to do the work of about 200 cubic feet of gas.

A very important factor which enters into the question of the cost of incandescent lighting is the care with which the arrangements of the electric light "wiring" have

---

* See *The Electrician*, March 16, 1894, p. 561.

been carried out. The "wiring" may be done in such a way that the user can only turn on and off large groups of lamps at once. As far as possible each lamp should be on a separate switch, and be capable of being used and put out of use alone. Such independent wiring costs more per light than group lighting, but it enables the user to create the necessary illumination with a minimum of expenditure of electric energy.

This is the moment to offer a word of caution against all and every form of "cheap" electric lighting work. Too much care cannot possibly be bestowed upon the quality of the materials and character of the work in laying the electric service lines and mains in buildings. This "wiring" has now become a special trade, and in the hands of competent firms is carried out with the utmost care and with all the numerous precautions which long experience has shown to be necessary for safety and efficiency. No amount of inspection after the work is finished and no amount of electric testing is really an efficient substitute for a watchful and experienced eye during the progress of the work. All important electric wiring work ought to be thus watched on behalf of the owners by a clerk of the works trained in this kind of supervision. It is not sufficient for a consulting engineer to draft the specification of the work; it must be suitably supervised during its progress. It is advisable that every architect should thus be competent to draft his own specification for electric housewiring, and that every clerk of the works should be competent to supervise and inspect the electric wiring work during its entire progress.

The actual average annual consumption of electric energy by incandescent lamps placed in buildings of various classes differs very greatly. Taking the case of supply by meter, it is found that for a large class of private houses the average number of

Board of Trade units taken per annum by an 8-c.p. lamp will not exceed 12 to 20. Indeed 20 is generally an outside limit. Lamps in clubs, hotels and restaurants show a very much larger average than 20, and may reach 30 to 40 B.T.U. per 8-c.p. lamp per annum, or occasionally even more. It may be mentioned that a 16-c.p. lamp would, under the same conditions, take twice, and a 32-c.p. three times the above amount. Since a 35-watt 8-c.p. lamp takes one B.T.U. of electric energy in 30 hours of continuous use, it follows that from 360 to 600 hours is the average total time during which a lamp in a private house is in use in the year. That is, from one hour to one hour and a half is the average daily time during which each lamp is alight in residential buildings. During the other 23 or $22\frac{1}{2}$ hours of the day, however, the electric supply station has to keep in readiness, machinery or other apparatus capable of giving current for that lamp should the user require it. The *load factor* of a supply station is the ratio of the actual amount of Board of Trade units sold to the amount which could be sold if the demand were constant and equal to the maximum demand. It will be seen from the above figures that the load factor of an electric supply station, supplying a purely residential district, does not exceed 8 or 10 per cent. The load factors of various classes of buildings are very different. The load factor of private houses is seen from the above figures to be even below 8 per cent. The most profitable consumers, therefore, from the point of view of the suppliers of electric current, are not those who have most lamps, but those who use the lamps they have for the longest number of hours.

We will now turn to consider the manner in which incandescent lamps can be employed in order to produce the best illuminating effect. The immense advantage which electric incandescent lighting possesses is, that it lends itself so readily to decorative purposes. In order that it may be employed in this sense to the best advantage, certain general guiding principles have to be held in view and followed. Unfortunately, this is not always done, and the result is that the effectiveness of the light, from the artistic point of view, is greatly diminished.

The first principle which should be allowed to guide us in the production of an artistic effect by electric lighting is that, while a proper and sufficient illumination is thrown upon the surfaces to be illuminated, the source of light must be itself, as far as possible, concealed. The illuminating power of a light for the purpose of vision does not depend merely upon the candle-power of the lamp, but it depends upon the amount of light which is received by the eye from the surfaces from which it is reflected. The pupil of the eye is capable of being varied over wide limits, probably from about four square millimetres, or the 150th part of a square inch, to 70 square millimetres or the tenth part of a square inch. This expansion or contraction of the pupil of the eye is effected by muscles over which the will has no control, and is termed in physiology a "reflex action." If the eye has been kept in darkness for some time, then the pupil becomes expanded to its greatest extent. Under these conditions, if the eye is opened, and especially if it is exposed to a bright light, this light falling on the retina causes an immediate muscular contraction of the pupil to be effected. It is very easy to see this adjusting process going on in the eye by shutting the eyes for a few minutes and then opening them in a bright light whilst a hand mirror is held before the face so as to examine the pupils of the eyes. The pupils will then be seen to be rapidly contracting in area immediately the eyes are opened. Hence, if the eye is turned upon a brilliant line of light, the contraction of the pupil which immediately sets in imposes a limitation upon the amount of light which can enter the eye; and if the eye is turned immediately towards a dull, badly illuminated surface, the pupil is not at once suddenly expanded again.

One part, at any rate, of the injurious effect produced by trying to read or write in the twilight is due to the strain produced in the eyes by the muscular effort thus demanded in the eye. When, therefore, we enter a room in which there are a number of bright lights, such, for instance, as a room in which incandescent lamps are being employed, the filaments of which can be seen, the moment that the eye is turned upon these bright lines of light, muscular contraction of the pupil sets in, and on turning the eye away again to other less illuminated surfaces, the retina of the eye does not receive sufficient light to observe details. In common language, the eye is "dazzled." There is a certain fascination about brilliant points or lines of light which captivates and attracts the eye even against the will.

It will thus be seen that the presence of brilliant lines or points of light in a room has a tendency to keep the pupil of the eye in a state of partial contraction, which unfits it for obtaining the best visual effect when turned upon surfaces less illuminated. Some eyes are especially sensitive in this respect. If, however, the lamp globes are made of frosted glass, or in other ways protected, so that the image of the incandescent filament cannot be directly thrown upon the retina, the actual visual effect may be increased in spite of the fact that such frosted globes or screens may cut off from 30 to 50 per cent. of the light. This is a common experience with everybody who tries to read or write in the neighbourhood of a brilliant incandescent lamp. By covering the globe with tissue paper, or with a ground glass or porcelain shade, although a diminution to a large extent is caused in the total light actually distributed from the lamp, yet, nevertheless, we are able to see better and more comfortably by it. The process, however, which is commonly adopted of cutting off a large proportion of the light by a semi-opaque globe is a wasteful one, because there can be no object in making light merely to absorb it.

The proper method is to so place the incandescent lamp that no light from the filament can directly enter the eye, but so that all the light sent out from the filament shall be reflected to the eye from the objects to be seen, such as the walls of the room and the various objects in the room,

FIG. 36.—View of the Interior of a Dining Room, Illuminated with Shaded Electric Candles.

and that the light shall reach the eye only after reflection from these surfaces. Bare or uncovered lamps, and especially if suspended half way down a room or on a level with, or a little above the eye, are a crude and exceedingly disagreeable method of illumination.

In all cases where incandescent lighting is carried out by those who understand the proper methods of using it, so as to produce the best artistic effects, the guiding principle is followed of so placing the lamps that the whole of the light emitted from them only reaches the eye after reflection from surrounding objects. This result can be attained in a variety of ways. The lamps may be placed in shades so as to throw the light upwards upon the ceiling of the room or upon the walls, from which surfaces it is ultimately reflected down on to the objects in the room. Wherever highly decorated and light ceilings exist in a room to be lit, this is a very effective method of distributing the light. In Figs. 36, 37, 38, 39, and 40, are shown photographs of interiors so illuminated. It is always possible to arrange glow lamps protecting them by metallic or other ornamental shields, so so as to obtain this desired effect. A natural shell, properly supported, forms a very effective screen and reflecting surface. The white pearly interior of the shell acts as a reflecting surface, and if the shell is very slightly translucent, then a pleasing effect is produced by the small and diffused light which passes through.

The second important principle of decorative electric lighting is to distribute the light properly and to prevent the concentration of light in large masses, thereby avoiding the production of harsh shadows. Everyone is familiar with the exceedingly disagreeable effect which is produced by sharp and strong shadows. A soft gradation of light and shade is essential in the production of an artistic effect in a room. The inartistic productions of many amateur photographers, especially in the region of portraiture, are much more due to the fact that they have not the means at their disposal for producing the right effects of light and shade on their subjects, than to any defects in their photographic manipulation. Any concentration of the light in large masses, in public or private rooms,

100 *ELECTRIC LAMPS AND ELECTRIC LIGHTING.*

Fig. 37.—View of the Interior of a Drawing Room Lighted Electrically with Screened Lamps.

Fig. 38.—View of a Ball Room Illuminated with Shaded Electric Candles in Chandeliers.

ELECTRIC GLOW LAMPS. 101

Fig. 39.—View of a Reception Room Illuminated with Inverted Incandescent Lamps throwing Light upwards.

Fig. 40.—View of a Drawing Room Illuminated with Shaded Electric Candles.

and especially when thrown downwards, invariably produces disagreeable shadows upon the face; and, therefore, in places like ball-rooms or drawing-rooms no such process of employing downwardly-suspended incandescent lamps should ever be permitted if it is desired to obtain the best results. All artists know the immense importance of obtaining proper lighting by diffused light, and the northern aspect which they invariably select for their studios is determined by this consideration. It is essential that the light should be so distributed in small units, in order that there shall be no sharp shadows cast upon any object. A simple test for the effectiveness of the distribution of light in any room may be obtained as follows: Take a white card or a sheet of white writing paper and hold it horizontally about the level of the eyes, then hold a pencil or other small rod vertically on the card, and note if any marked shadow of the pencil is thrown in any direction: if it is, then the light is insufficiently diffused, and the lamps ought to be re-arranged. The best effects as to distribution of light are generally obtained by the employment of 5 candle-power lamps. These glow lamps are constructed with bulbs of semi-opaque glass having the shape of a candle flame. (see Fig. 21). These candle lamps are carried on the extremity of an artificial porcelain candle, so as to resemble in some degree an ordinary wax candle alight. These lamps are then arranged in groups on chandeliers or brackets, being protected in many cases from direct visibility by plain or ornamental shades. In Figs. 36 and 41 are shown photographic illustrations of rooms which are lit by such candle lamps, and in which the light from these is thrown up against the walls of the room, the candle lamps not being themselves directly visible.

A third great guiding principle of artistic lighting is to proportion the light to the nature of the surfaces from which it is reflected. We require light in our living rooms to see the

ELECTRIC GLOW LAMPS. 103

Fig. 41.—Drawing Room Illuminated with Electric Lamps, so shaded that no direct view of the lamp filament can be obtained.

Fig. 42.—Interior of Freemason's Hall Illuminated by Shaded Incandescent Lamps on Frieze and Ceiling.

pictures, decorations, objects, and, above all, to see one another; and we can only see these objects by the light which they reflect back to the eye. The aim should, therefore, be to have sufficient but not excessive incident light. It should be borne in mind that the different surfaces with which the walls of rooms are covered have very different reflective powers. Dull walls absorb eighty per cent. of the light falling on them, and reflect only about twenty per cent. of the light; this is the case, for instance, with walls covered with dark oak panelling or dull-coloured paint or paper. Ordinary light tints of paper or paint may reflect from forty to sixty per cent.; clean white surfaces, such as white painted and varnished wood or plaster, may reflect as much as eighty per cent., and mirrors from eighty to ninety per cent.

The following figures have been given by Dr. Sumpner for the reflecting power of various surfaces:—

Table *of Reflecting Powers of Various Surfaces.*

White blotting paper 82 per cent.	Plane deal (clean) 40–50 per cent.
Ordinary foolscap ... 70 ,, ,,	Plane deal (dirty) ... 20 ,, ,,
Newspapers.........50–70 ,, ,,	Yellow painted wall
Yellow wall paper ... 40 ,, ,,	(dirty) .... 20 ,, ,,
Blue paper .... 25 ,, ,,	Yellow painted wall
Dark brown paper ... 13 ,, ,,	(clean) .... 40 ,, ,,
Deep chocolate paper 4 ,, ,,	Black cloth .... 1·2 ,, ,,
	Black velvet .... ·4 ,, ,,

Everyone is familiar with the way in which the laying of a white table-cloth brightens up a dark dining room, or in which the substitution of white for dark curtains lightens up a room. The reason for this is obvious from the above table. Hence, rooms with dark oil paintings, dark woodwork, hangings, curtains, or portières require relatively much more illumination per square foot than rooms with very light decorations, water-colour paintings, mirrors, &c. Experience has shown that in a room of the latter description an illumination is required which is equal to an expenditure in the lamp of about $\frac{3}{4}$ or 1 watt per square foot of floor surface, on the assumption that the lamps are placed, on an average, about 8 or 9 feet above the floor. A room of the former description, such as a picture gallery with dark oak walls, may require from 2 to 3 watts per square foot of floor surface. These figures, however, cannot be taken too absolutely. Experience is the only guide as to the amount of light to be placed in a room. As a broad general rule, 100 square feet of floor surface will be barely illuminated by one 16-c.p. lamp placed about 8 feet above the floor. It will be well illuminated by two such lamps placed 8 feet above the floor, and brilliantly illuminated by four such lamps. In other words, one 16-c.p. lamp per 100 square feet of floor surface is hardly sufficient, one 16-c.p. lamp per 50 square feet of floor surface is fairly good, and one for every 25 square feet of floor surface is brilliant. These figures are given on the assumption that the lamp is placed about 8 feet above the floor, and that the floor surface is fairly reflective. The lighting of pictures is a very special study, and can only be properly undertaken by those who have given considerable attention to the method of placing and using incandescent lamps. The mere illumination of a room by a number of lamps hung head downwards by cord pendants from the ceiling, a method which is not unusually adopted, is the most crude and imperfect method of applying the electric light, and ought never to be permitted in any case where the smallest decorative effect is really desired.

The brief limits of a lecture will not permit us to enlarge at greater length on this part of the subject, but it opens up a wide field for the exercise of the inventive and æsthetic faculties.

Before leaving the subject of incandescent lamps it will not be without interest to refer briefly to the investigations which have been made into a curious effect, found to exist in lamps which have a third or idle metallic wire sealed through the glass. Our starting-point for this purpose is a discovery made by Mr. Edison in 1884, and which received examination at the hands of Sir W. H. Preece in the following year,* and by the author more recently. Here is the initial experiment. A glow-lamp having the usual horseshoe-shaped carbon (Fig. 43) has a metal plate held on a platinum wire sealed through the glass bulb. This plate is so fixed that it stands up between the two sides of the carbon arch without touching either of them. We shall illuminate the lamp by a

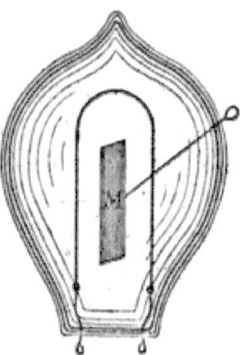

Fig. 43.—Glow Lamp having insulated metal middle plate M sealed into the bulb to exhibit the "Edison effect."

continuous current of electricity, and for brevity's sake speak of that half of the loop of carbon on the side by which the current enters it as the positive leg, and the other half of the

---

* Sir W. H. Preece's interesting Paper on this subject is published in the *Proceedings of the Royal Society* for 1885, p. 219. See also *The Electrician*, April 4, 1885, p. 436, "On a Peculiar Behaviour of Glow-lamps when Raised to High Incandescence." The following statements of experiments are chiefly taken from a Friday evening discourse given by the author at the Royal Institution, February 14, 1890, on "Problems in the Physics of an Electric Lamp," and from a Paper on "Electric Discharge between Electrodes at different Temperatures in Air and in High Vacua." (*Proceedings of the Royal Society, 1890.*)

loop as the negative leg. The diagram in Fig. 45 shows the position of the plate with respect to the carbon loop. There is a distance of half an-inch, or in some cases many inches, between either leg of the carbon and this middle plate. Setting the lamp in action, I connect a sensitive galvanometer between the middle plate and the *negative terminal* of the lamp, and you see that there is no current passing through the instrument. If, however, I connect the terminals of my galvanometer to the middle plate, and to the *positive electrode* of the lamp, we find a current of some milliamperes is passing through it. The diagrams in Fig. 45 show the mode of con-

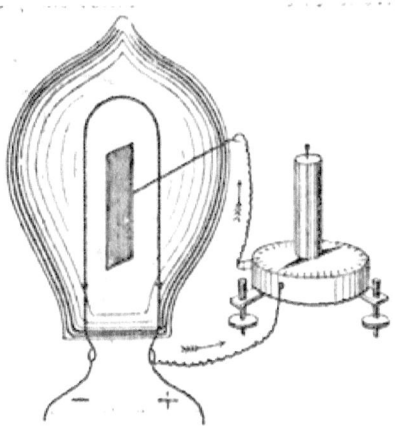

Fig. 44.—Sensitive Galvanometer connected between the middle plate and positive electrode of a Glow Lamp, showing current flowing through it when the lamp is in action ("Edison effect").

nection of the galvanometer in the two cases. This effect, which is often spoken of as the "Edison effect," clearly indicates that an insulated plate so placed in the vacuum of a lamp in action is brought down to the same potential or electrical state as the negative electrode of the carbon loop. On examining the direction of the current through the galvanometer we find that it is equivalent to a flow of negative electricity taking place through it *from* the middle plate *to* the positive electrode of the lamp. A consideration cf this fact shows us that there must be some way by which negative electricity gets across the vacuous space from the negative leg of the carbon to the metal plate, whilst at the same time a negative charge cannot pass from the metal plate across to the positive leg.

Before passing away from this initial experiment, the attention of the reader should be called to a curious effect at the moment when the lamp is extinguished. Connecting the galvanometer, as at first, between the middle plate and the negative electrode of the lamp, we notice that, though made

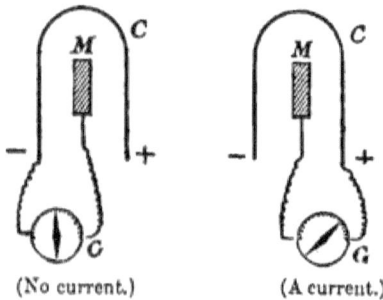

Fig. 45.—Mode of connection of galvanometer G to middle plate M and carbon horseshoe-shaped conductor C in the experiment of the "Edison effect."

highly sensitive, the galvanometer indicates no current flowing through it whilst the lamp is in action. Switching off the current from the lamp produces, as you see, a violent kick or deflection of the galvanometer, indicating a sudden rush of current through it.

In endeavouring to ascertain further facts about this effect one of the experiments which early suggested itself was directed to determine the relative effects of different portions of the carbon conductor. Here is a lamp (*see* Fig 46) in which one leg of the carbon horseshoe has been enclosed in a glass tube the size of a quill, which shuts in one half of the carbon. The bulb contains, as before, an insulated middle plate. If we pass the actuating current through this lamp in such a direction that the covered or sheathed leg is the *positive* leg, we find the effect existing as before. A galvanometer connected between the plate and positive terminal of the lamp yields a strong current, whilst if connected between the negative terminal and the middle plate there is no current at all. Let us, however, reverse the current through the lamp so that the

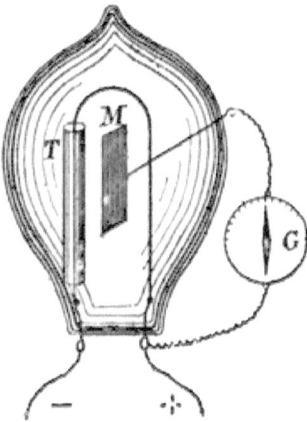

Fig. 46.—Glow Lamp having negative leg of carbon enclosed in glass tube T, the "Edison effect" being thereby annulled or greatly diminished.

shielded or enclosed leg is now the negative one, and the galvanometer is able to detect no current, whether connected in one way or in the other. We establish, therefore, the conclusion that it is the negative leg of the carbon loop which is the active agent in the production of this "Edison effect," and that, if it is enclosed in a tube of either glass or metal, no current is found flowing in a galvanometer connected between the positive terminal of the lamp and this middle collecting plate.

Another experiment which confirms this view is as follows: The lamp (Fig. 47) has a middle plate, which is provided with a little mica flap or shutter on one side of it. When the lamp is held upright the mica shield falls over and covers one side of the plate, but when it is held in a horizontal position the mica shield falls away from the front of the plate and exposes it. Using this lamp as before, we find that, when the positive leg of the carbon loop is opposite to the shielded face of the plate, we get the "Edison effect" as before in any position of the lamp. Reversing the lamp current, and making

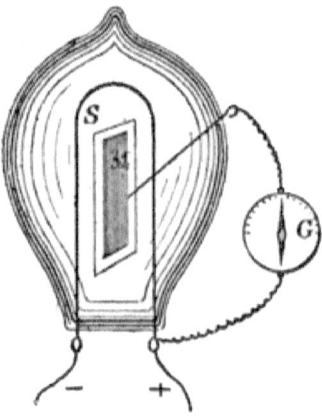

Fig. 47. — Glow Lamp having mica shield S interposable between middle plate M and negative leg of carbon, thereby diminishing the "Edison effect."

that same leg the *negative* one, we find that, when the lamp is so held, the metal plate is shielded by the interposition of the mica, and the galvanometer current is very much less than when the shield is shaken on one side and the plate exposed fully to the negative leg

At this stage it will perhaps be most convenient to outline briefly the beginnings of a theory which may be proposed to reconcile these facts, and leave you to judge how far the subsequent experiments seem to confirm this hypothesis. Very briefly, this theory is as follows: From all parts of the incandescent carbon loop, but chiefly from the negative leg, carbon molecules are being projected which carry with them, or are charged with, negative electricity. In a few moments a suggestion will be made to you which may point to a possible hypothesis on the manner in which the molecules acquire this negative charge. Supposing this, however, to be the case, and that the bulb is filled with these negatively-charged molecules, what would be the result of introducing into their midst a conductor such as this middle metal plate which is charged positively? Obviously, they would all be attracted to it and discharge against it. Suppose the positive charge of this conductor to be continually renewed, and the negatively-charged molecules continually supplied—which conditions can be obtained by connecting the middle plate to the positive electrode of the lamp—the obvious result will be to produce a current of electricity flowing through the wire or galvanometer by means of which this middle plate is connected to the positive electrode of the lamp. If, however, the middle plate is connected to the negative electrode of the lamp, the negatively-charged molecules can give up no charge to it, and produce no current in the interpolated galvanometer. We see that, on this assumption, the effect must necessarily be diminished by any arrangement which prevents these negatively-charged molecules from being shot off the negative leg or from striking against the middle plate.

Another obvious corollary from this theory is that the "Edison effect" should be annihilated if the metal collecting plate is placed at a distance from the negative leg much greater than the mean free path of the molecules.

Here are some experiments which confirm this deduction. In this bulb (Fig. 48) the metal collecting plate, which is to be connected through the galvanometer with the positive terminal of the lamp, is placed at the end of a long tube opening out of and forming part of the bulb. We find the "Edison effect" is entirely absent, and that the galvanometer current is zero. We have, as it were, placed our target at such a distance that the lowest range molecular bullets cannot hit it, or at least that but very few of them do so. Here, again, is a lamp in which the plate is placed at the extremity of a tube opening out of the bulb, but bent at right angles (Fig. 49). We find in this case, as first discovered by Mr. W. H. Preece, that there is no "Edison effect." Our molecular marksman

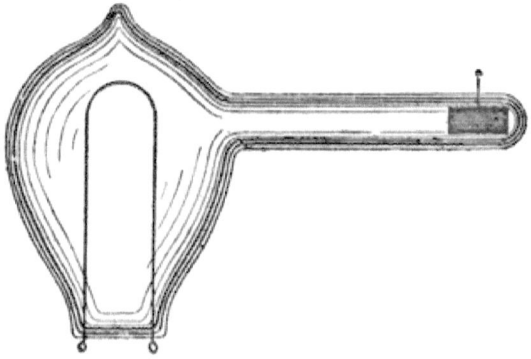

Fig. 48.—Collecting plate placed at end of tube, 18in. in length, opening out of bulb.

cannot shoot round a corner. None of the negatively-charged molecules can reach the plate, although that plate is placed at a distance not greater than would suffice to produce the effect if the bend were straightened out. Following out our hypothesis into its consequences would lead us to conclude that the material of which the plate is made is without influence on the result, and this is found to be the case.

We should expect also to find that the larger we make our plate, and the nearer we bring it to the negative leg of the carbon, the greater will be the current produced in a circuit connecting this plate to the positive terminal of the lamp. I have before me a lamp with a large plate placed very near the negative leg of the carbon of a lamp, and we find that we can collect enough current from these molecular charges to work a telegraph relay and ring an electric bell. The current which is now working this relay is made up of the charges collected by the plate from the negatively-charged carbon molecules, which are projected against it from the negative leg, across the highly perfect vacuum. I have tried experiments with lamps in which the collecting plate is placed in all kinds

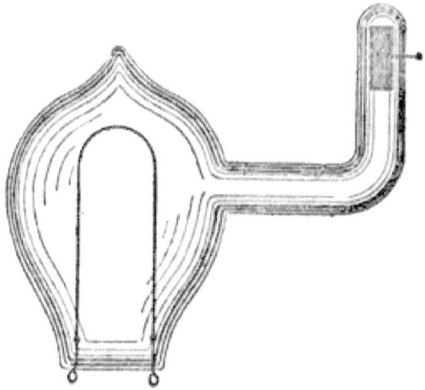

FIG. 49.—Collecting plate placed at end of elbow tube opening out of bulb.

of positions, but the results may all be summed up by saying that the greatest effects are produced when the collecting plate is as near as possible to the base of the negative end of the loops, and, as far as possible, encloses, without touching, the carbon conductor. It is not necessary to make more than a passing reference to the fact that the magnitude of the current flowing through the galvanometer when connected between the middle plate and the positive terminal of the lamp often "jumps" from a low to a high value, or *vice versâ*, in a remarkable manner, and that this sudden change in the current can be produced by bringing strong magnets near the outside of the bulb.

Let us now follow out into some other consequences the hypothesis that the interior of the bulb of a glow lamp, when in action, is populated by flying crowds of carbon atoms all carrying a negative charge of electricity. Suppose we connect our middle collecting plate with some external reservoir of electric energy, such as a Leyden jar, or with a condenser equivalent in capacity to many hundreds of Leyden jars, and let the side of the condenser which is charged positively be first placed in connection through a galvanometer with the middle plate (*see* Fig. 50), whilst the negative side is placed

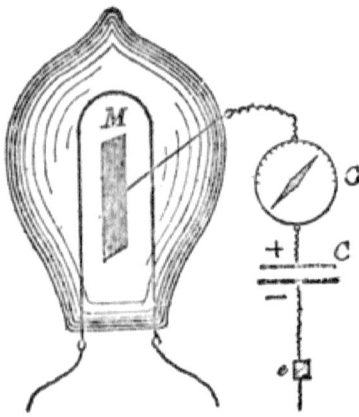

Fig. 50.—Charged condenser C discharged by middle plate M, when the positively-charged side of condenser is in connection with the plate and the other side to earth e.

in connection with the earth. Here is a condenser of two microfarads capacity so charged and connected. Note what happens when I complete the circuit, and illuminate the lamp by passing the current through its filament. The condenser is at once discharged. If, however, we repeat the same experiment, with the sole difference that the negatively-charged side of the condenser is in connection with the middle plate, then there is no discharge.

These very interesting experimental results may also be regarded from another point of view. In order that the condenser may be discharged, as in the first case, it is essential that the negatively-charged side of the condenser shall be in connection with some part of the circuit of the incandescent carbon loop. This experiment with the condenser discharged by the lamp may be, then, looked upon as an arrangement in which the plates of a charged condenser are connected respectively to an incandescent carbon loop and to a cool metal plate, both being enclosed in a highly-vacuous space; and it appears that the discharge takes place when the incandescent conductor is the negative electrode of this arrangement, but not when the cooler metal plate is the negative electrode of the charged condenser. The negative charge of the condenser can be carried across the vacuous space from the hot carbon to the colder metal plate, but not in the reverse direction.

This experimental result led me to examine the condition of the vacuous space between the middle metal-plate and the negative leg of the carbon loop in the case of the lamp employed in our first experiment. Let us return for a moment to that lamp. I join the galvanometer between the middle plate and the negative terminal of the lamp, and find, as before, no indication of a current. The metal plate and the negative terminal of the lamp are at the same electrical potential. In the circuit of the galvanometer we will insert a single galvanic cell, having an electromotive force of rather over one volt. In the first place let that cell be so inserted that its negative pole is in connection with the middle plate, and its positive pole in connection through the galvanometer with the negative terminal of the lamp (*see* Fig. 51). Regarding the circuit of that cell alone, we find that it consists of the cell itself, the galvanometer wire, and that half-inch of highly vacuous space between the hot carbon conductor and the middle plate. In that circuit the cell cannot send any sensible current at all, as it is at the present moment connected up. But if we reverse the direction of the cell, so that its positive pole is in connection with the middle plate, the galvanometer at once gives indications of a very sensible current. This highly-vacuous space, lying between the middle metal plate on the one hand and the incandescent carbon on the other, possesses a kind of unilateral conductivity, in that it will allow the current from a single galvanic cell to pass one way but not the other. It is a very old and familiar fact, that in order to send a current from a battery through a highly-rarefied gas by means of metal electrodes

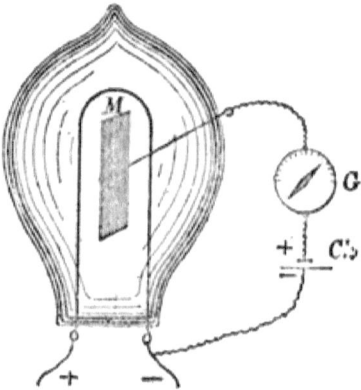

FIG. 51.—Current from Clark cell Ck being sent across vacuous space between negative leg of carbon and middle plate M. Positive pole of cell in connection with plate M through galvanometer G.

the electromotive force of the battery must exceed a certain value. Here, however, we have indication that if the negative electrode by which that current seeks to enter the vacuous space is made incandescent the current will pass at a very much lower electromotive force than if the electrode is not so heated.

A little consideration of the foregoing experiments led to the conclusion that, in the original experiment, as devised by Mr.

Edison, if we could by any means render the middle plate very hot, we should get a current flowing through a galvanometer when it is connected between the middle plate and the negative electrode of the carbon. This experiment can be tried in the manner now to be shown. Here is a bulb (Fig. 52) having in it two carbon loops: one of these is of ordinary size, and will be rendered incandescent by the current from the mains. The other loop is very small, and will be heated by a well-insulated secondary battery. This smaller incandescent loop shall be employed just as if it were a middle metal plate. It is, in fact, simply an incandescent middle conductor. On repeating

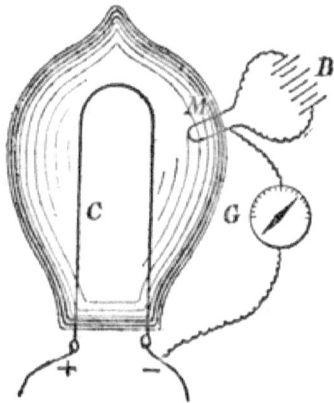

Fig. 52.—Experiment showing that when the "middle plate" is a carbon loop rendered incandescent by insulated battery B, a current of negative electricity flows from M to the positive leg of main carbon C across the vacuum.

the typical experiment with this arrangement, we find that the galvanometer indicates a current when connected between the middle loop and either the positive *or* the negative terminal of the main carbon. I have little doubt but that, if we could render the platinum plate in our first-used lamp incandescent by concentrating on it from outside a powerful beam of radiant heat, we should get the same result.

A similar set of results can be arrived at by experiments with a bulb constructed like an ordinary vacuum tube, and having small carbon loops at each end instead of the usual platinum or aluminium wires. Such a tube is now before you (*see* Fig. 53), and will not allow the current from a few cells of a secondary battery to pass through it when the carbon loops are cold. If, however, by means of well-insulated secondary batteries, we render both of the carbon loop electrodes highly incandescent, a single cell of a battery is sufficient to pass a very considerable current across that vacuous space, provided the resistance of the rest of the circuit is not large. We may embrace the foregoing facts by saying that if the electrodes, but especially the negative electrode, which form the means of ingress and egress of a

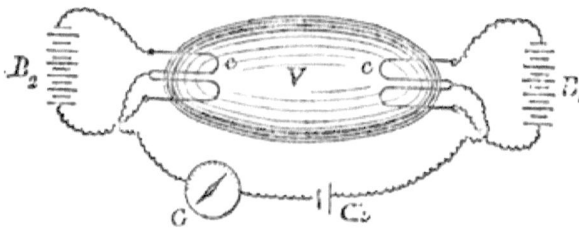

Fig. 53.—Vacuum tube having carbon loop electrodes, *c c*, at each end, rendered incandescent by insulated batteries, $B_1 B_2$, showing current from Clark cell, *Ck*, passing through the high vacuum when the electrodes are incandescent.

current into a vacuous space, are capable of being rendered highly incandescent, and if at that high temperature they are made to differ in electrical potential by the application of a very small electromotive force, we may, under these circumstances, get a very sensible current through the rarefied gas. If the electrodes are cold, a very much higher electromotive force will be necessary to get the discharge or current through the space. These facts have been made the subject of elaborate investigation by Hittorf, Goldstein, and more recently by J. J. Thompson and others.

The subject of electrical discharge through rarefied air or gases has attracted the attention of many experimentalists. The outcome of their work has been to indicate that when two electrodes are placed in a glass vessel containing a gas under very reduced pressure, the action of electromotive force is to create two streams of particles which may be either chemical atoms or particles smaller than chemical atoms which carry positive electricity in one direction and negative in the other. It may also be that particles of the electrodes are torn off and take part in this convective process. It is tolerably certain that this last effect does occur in the case of the carbon filament electric lamp.

It has been found that a clean middle plate placed between the legs of the carbon horseshoe becomes black soonest on the side facing the negative leg of the carbon. It has also been noticed that, in the case of lamps with treated filaments, the surface of the deposit assumes a dead lamp-black appearance soonest on the negative leg. These observations tend to support the contention that there is an ejection, or more rapid ejection, of carbon from the negative leg of the carbon horseshoe. Much has yet to be done before the full meaning of this "Edison effect" is unravelled; but, as far as it has yet been examined, the following summary of facts includes most of those as yet observed.

If a platinum wire is sealed through the glass bulb of an ordinary carbon filament lamp, and carries at its extremity a metal plate so placed as to stand up between the legs of the carbon horseshoe without touching either of them, then, when the lamp is actuated by a continuous current, it is found that:—

(1.) This insulated metal plate is brought down instantly to the potential of the base of the negative leg of the carbon, and no sensible potential difference exists between the insulated metal plate and the negative electrode of the lamps, whether the test be made by a galvanometer, by an electrostatic voltmeter, or by a condenser.

(2.) The potential difference of the plate and the positive electrode of the lamp is exactly the same as the working potential difference of the lamp electrodes, provided this is measured electrostatically, *i.e.*, by a condenser, or by an electrostatic voltmeter taking no current; but, if measured by a galvanometer, the potential difference of the plate and the positive electrode of the lamp is something less than that of the working lamp electrodes.

(3.) This absolute equality of potential between the negative electrode of the lamp and the insulated plate only exists when the carbon filament is in a state of vivid incandescence, and when the insulated plate is not more than an inch or so from the base of the negative leg. When the lamp is at intermediate stages of incandescence, or the plate is considerably removed from the base of the negative leg, then the plate is not brought down quite to the same potential as the negative electrode.

(4.) A galvanometer connected between the insulated plate and the *positive* electrode of the lamp shows a current increasing from zero to four or five milliamperes, as the carbon is raised to its state of commercial incandescence. There is not any current greater than 0·0001 of a milliampere* between the plate and the *negative* electrode when the lamp has a good vacuum.

(5.) If the lamp has a bad vacuum this inequality is destroyed, and a sensitive galvanometer shows a current flowing through it when connected between the middle plate and either the positive *or* negative electrode.

(6.) When the lamp is actuated by an *alternating* current, a *continuous* current is found flowing through a galvanometer

---

* A milliampere is the $\frac{1}{1000}$th of an ampere.

connected between the insulated plate and *either* terminal of the lamp. The direction of the current through the galvanometer is such as to show that negative electricity is flowing from the plate through the galvanometer to the lamp terminal. This is also the case in (4); but, if the lamp has a bad vacuum, then negative electricity flows *from* the plate through the galvanometer *to* the positive terminal of the lamp, and negative electricity flows *to* the plate through the galvanometer *from* the negative terminal of the lamp.

(7.) The same effects exist, on a reduced scale, when the incandescent conductor is a platinum wire instead of a carbon filament. The platinum wire has to be brought up very near to its point of fusion in order to detect the effect, but it is found that a current flows between the positive electrode of a platinum wire lamp and a platinum plate placed in the vacuum near to the negative end of that wire.

(8.) The material of which the plate is made is without influence on the result. Platinum, aluminium, and carbon have been indifferently employed.

(9.) The active agent in producing this effect is the *negative* leg of the carbon. If the negative leg of the carbon is covered up by enclosing it in a glass tube, this procedure entirely, or nearly entirely, prevents the production of a current in a galvanometer connected between the middle plate and the positive terminal of the lamp.

(10.) It is a matter of indifference whether a glass or metal tube is employed to cover up the negative leg of the carbon; in any case this shielding destroys the effect.

(11.) If, instead of shielding the negative leg of the carbon, a mica screen is interposed between the negative leg and the side of the middle plate which faces it, then the current produced in a galvanometer connected between the positive terminal of the lamp and the middle plate is much reduced. Under the same circumstances hardly any effect is produced when the mica screen is interposed on that side of the metal plate which faces the positive leg of the carbon.

(12.) The position of the metal plate has a great influence on the magnitude of the current traversing a galvanometer connected between the metal plate and the positive terminal of the lamp. The current is greatest when the insulated metal plate is as near as possible to the base of the negative leg of the carbon, and greatest of all when it is formed into a cylinder which embraces, without touching, the base of the negative leg. The current becomes very small when the insulated metal plate is removed to 4 or 5 inches from the negative leg, and becomes practically zero when the metal plate is at the end of a tube forming part of the bulb, which tube has a bend at right angles in it. Copious experiments have been made with metal plates in all kinds of positions.

(13.) The galvanometer current is greatly influenced by the surface of the metal plate: being greatly reduced when the surface of the plate is made small, or when the plate is set edgeways to the negative leg, so as to present a very small apparent surface when seen from the negative leg. In a lamp having the usual commercial vacuum, the effect is extremely small when the insulated metal plate is placed at a distance of 18in. from the negative leg, but even then it is just sensible to a very sensitive galvanometer.

(14.) If a charged condenser has one plate connected to the insulated metal plate, and the other plate connected to any point of the circuit of the incandescent filament, this condenser is instantly discharged if the positively-charged side of the condenser is connected to the insulated plate and the negative side to the hot filament. If, however, the negative leg of the carbon horseshoe is shielded by a glass tube, this discharging power is much reduced, or altogether removed.

(15.) If the middle plate consists of a separate carbon loop, which can itself be made incandescent by a separate insulated battery, then, when this middle carbon is rendered incandescent, and is employed as the metal plate in the above experiment, the condenser is discharged when the negatively-charged side of it is connected to the hot middle carbon, the positively-charged side of it being in connection with the principal carbon horseshoe.

(16.) If this last form of lamp is employed as in (4), the subsidiary carbon loop being used as a middle plate, and a galvanometer being connected between it and either the positive or negative main terminal of the lamp, then, when the subsidiary carbon loop is cold, we get a current through the galvanometer only when it is in connection with the positive main terminal of the lamp; but, when the subsidiary carbon is made incandescent by a separate insulated battery, we get a current through the galvanometer when it is connected either to the positive *or* to the negative terminal of the lamp. In the first case the current through the galvanometer is a negative current flowing from the middle carbon to the positive main terminal, and in the second case it is a negative current from the negative main terminal to the middle subsidiary hot carbon.

(17.) If a lamp having a metal middle plate held between the legs of the carbon loop has a galvanometer connected between the negative main terminal of the lamp and this middle plate, we find that, when the carbon is incandescent, there is no sensible current flowing through the galvanometer. The vacuous space between the middle plate and the hot negative leg of the carbon possesses, however, a curious unilateral conductivity. If a single galvanic cell is inserted in series with the galvanometer, we find that this cell can send a current deflecting the galvanometer when its negative pole is in connection with the negative main terminal of the lamp, but if its positive pole is in connection with the negative terminal of the lamp, then no current flows. The cell is thus able to force a current through the vacuous space when the direction of connection of the cell is such as to cause negative electricity to flow across the vacuous space from the hot carbon to the cooler metal plate, but not in the reverse direction.

(18.) If a vacuum tube is constructed, having horseshoe carbon filaments sealed into it at each end, and which can each be made separately incandescent by an insulated battery, we find that such a vacuum tube, though requiring an electromotive force of many thousands of volts to force a current through it when the carbon loops are used as electrodes and are *cold*, will yet pass the current from a single galvanic cell when the carbon loop which forms the negative electrode is rendered incandescent. It is thus found that a high vacuum terminated electrically by unequally-heated carbon electrodes possesses a unilateral conductivity, and that electric discharge takes place freely through it under an electromotive force of a few volts when the *negative* electrode is made highly incandescent.

Some reference must be made before concluding this Lecture to the evolution of the *high-voltage incandescent lamp*, and the direction in which invention is tending with respect to the utilisation of high-working electromotive forces in electric lighting.

When considering the subject of electric distribution, in the last Lecture, it will be pointed out that there are strong arguments in favour of the use of electric pressures as high, or higher, than 200 volts in domestic and private electric lighting when conducted from public electric supply stations. Up to a few years ago the usual limit of electric pressure for working incandescent lamps was 110 or 120 volts at the terminals of the lamp. The clear recognition of the advantages to be gained by the employment of 200 or 250 volt lamps led to strong representations being made to the Board of Trade to induce them to raise the official limit defining the minimum pressure which might be introduced into the premises of a user of electric lamps from public electric supply stations. The then existing regulations were accordingly revised by the Board of Trade, and it is now ordered that an electric pressure, if continuous current, of less than 500 volts, and, if alternating current, of less than 250 volts, shall be considered as a *low electric pressure*. Systems of electric supply are, therefore, now very generally adopted in which the service pressure to the user is 200 or 250 volts. A demand accordingly sprung up for an incandescent lamp to work at the above pressures.

It must be remembered that whilst the type of incandescent electric lamp most largely in use since the commencement of electric lighting has probably been the 16 c.p. lamp, there is also a considerable use of 8 c.p. lamps, and also a considerable demand for 5 c.p. lamps. As long as we are limited to the use of carbon alone in the lamp filament it has not hitherto been found that lamps having an electrical power consumption at a less rate than 3 watts per candle-power have a sufficiently satisfactory duration on ordinary circuits to render them widely adopted at present. The reason for this is that, as the watts per candle are decreased, the temperature of the carbon rises, and a point is then reached at which volatilisation of the carbon proceeds very rapidly. Hence we may say that the normal type of 16 c.p. incandescent lamp will absorb at least 50 watts, and more usually 60 watts, whilst the 8 c.p. lamps and 5 c.p. lamps will take up about 30 watts and 18 to 20 watts respectively. If, then, we are working at an electric pressure of 100 volts, the table on the next page shows the currents taken by these lamps and their carbon filament resistances when hot—that is to say, measured when the lamp filament is incandescent. Many lamp manufacturers issue lamps of various "efficiencies" marked for voltages so that they take $2\frac{1}{2}$, 3, $3\frac{1}{2}$ or 4 watts per candle. Generally speaking, when used on ordinary public electric supply circuits, the lower the "watts per candle" the shorter the life of the lamp.

Currents and Hot Resistances of Carbon Filament Lamps working at 100 volts Pressure and 3·5 watts per Candle-power.

Candle-power of lamp.	Power taken up in lamp in watts.	Lamp current in amperes.	Hot resistance of filament in ohms.
16 c.p.	56·0	0·56	178
8 c.p.	28·0	0·28	356
5 c.p.	17·5	0·175	571

It will be seen, therefore, that the lower the candle-power of the 100-volt lamp the higher must be the resistance of the filament.

This necessitates a much thinner filament in the case of the lower candle-power lamps, assuming all to be made for working at the same voltage. The proposal, therefore, to use 200-volt lamps of 16, 8 or 5 c.p. involved the construction of lamps taking half the currents of the above-mentioned 16, 8 and 5 c.p. 100-volt lamps, and having filament resistances *four times* as great.

In the manufacture of carbon filaments for lamps, it is found that the resistivity, or specific resistance, of carbon is reduced by all processes which tend to make a very dense or tough form of carbon. Thus, for instance, the simple carbonisation of a very thin strip of bamboo (as in the old form of Edison lamp) yields a carbon which, when examined microscopically, is seen to be of a porous and cellular structure; in fact the structure of the vegetable fibre is retained after carbonisation to a considerable extent. A carbon of this kind has a relatively high specific resistance.

The carbon produced by carbonising a cotton thread which has been treated with dilute sulphuric acid to destroy its organic structure and produce a more homogeneous carbonisable structure has a lower specific resistance than the simply charred organic fibre. All modern carbon filament lamps are now made by the carbonisation of some form of soluble cellulose squirted or pressed out into a continuous thread resembling catgut. Cellulose is dissolved by some suitable solvent, such as zinc chloride solution, and the resulting gummy mass is pressed out through a die into a fine uniform thread or wire. This material, when given proper shape, is carbonised in a furnace out of contact with air. Carbons made in the above manner are called primary carbons.

It is usual, however, to subject the primary carbon to a process called "treating" or "flashing." The filament is placed in a vessel enclosing an atmosphere consisting of a hydrocarbon. The gas generally used is rarefied coal gas. If the carbon filament in this position is gradually raised to a red heat and finally to a bright white heat by passing an electric current through it, the surrounding hydrocarbon gas or vapour is decomposed and carbon is deposited on the filament. Moreover, if the primary filament is non-uniform, and has defects or places of high resistance, the current will produce heat in these places at the greatest rate, and will, therefore, raise them to the highest temperature. Accordingly, the carbon-depositing process takes effect at these places most quickly, and the tendency is, to some extent, to remedy defects in the primary carbon and to make the resulting filament more uniform in resistance. Moreover, by controlling the thickness of the deposit of secondary carbon, the lamp maker can adjust the resistances of filaments so that they shall be of exactly the required resistance.

This "deposited carbon" has a much denser structure than the primary carbon. When a "treated" filament is examined under the microscope and compared with an "untreated" filament it is seen that the surface of the treated filament has a bright steely lustre and has a very uniform and metallic surface. If the treating has been badly or hastily performed, or carried on too long, the surface will be nodulous or deformed by little warts or projections of deposited carbon. The secondary or deposited carbon has also, *per se*, a much lower specific resistance than any form of primary carbon produced by the simple carbonisation of an organic fibre. A primary carbon can, however, be prepared with a dense elastic structure by heating an organic fibre or carbonised fibre to an exceedingly high temperature in an electric furnace.

Generally speaking, the higher the temperature at which the carbonisation of organic material has been conducted the denser the form of carbon produced, and the lower its specific electrical resistance. At very high temperatures all forms of carbon pass into graphite, and the deposited carbon or electric furnace-produced carbon consists largely of carbon in the graphitic form. The three forms of carbon, viz., charcoal, graphite and diamond, have densities which are very different, as shown by the following table :—

Density or Specific Gravity of Carbon in Various Forms.
Water = 1.

Material.	Density.
Coke	about 0·4.
Hard charcoal	0·47 to 0·57.
Pit coal	1·20 ,, 1·50.
Carbonised cellulose	about 1·35.
Anthracite coal	1·41.
"Deposited" carbon	1·95 to 2·28.
Graphite	2·10 ,, 2·35.
Diamond	3·3 ,, 3·55.

It will be seen that the density of deposited carbon, or the carbon precipitated on a filament by heating it in coal gas or benzol vapour, approximates to that of graphite. The density of a simply carbonised thread or bamboo slip is similar to that of most varieties of coal.

The specific electric resistances of these different varieties of carbon vary very much. The diamond is an exceedingly good insulator. Coke and charcoal, in the more porous forms, are very poor conductors, but graphite has an electric conductivity which places it almost in the same category as metals.

In lamp manufacture it is usual to state the electric resistance of filament materials per *inch-mil*. The *mil* is the name for one-thousandth of an inch. If a very fine wire is made, of circular cross-section and one-thousandth of an inch in diameter, this is said to be one mil thick. If an inch in length of this wire is cut off, the piece would be called one inch-mil. The following table gives the electrical resistances per inch-mil of various kinds of carbon:—

Electrical Resistances in Ohms per inch-mil of Various Forms of Carbon.

Small arc lamp carbon rod	about 4,000 ohms
Jablochkoff candle rod	,, 2,000 ,,
Rod of Carré carbon 1mm. thick	,, 1,700 ,,
Carbonised bamboo	,, 3,000 ,,
Carbonised parchmentised thread	,, 2,000 to 2,500 ohms
Ordinary carbon filament, "treated" or flashed	,, 1,200 to 1,700 ,,
Deposited or secondary carbon	,, 300 to 450 ,,

The above values for the specific resistance of various carbons in ohms per inch-mil can be converted into microhms per cubic centimetre sufficiently nearly by multiplying by 2.

The resistance of all conductive forms of carbon diminishes as temperature increases. The very interesting observation has been recorded by Mr. J. W. Howell (*The Electrician*,

Vol. XXXVIII., p. 836) that "treated" carbon filaments have a minimum resistance corresponding to a certain temperature. He also found this to be true of graphite.

It will, therefore, be seen that the graphitic carbon deposited upon a primary carbon, when it is heated in coal gas or benzol, has a specific electrical resistance of about one-eighth to one-tenth that of simple carbonised organic material. Accordingly, if a filament is treated, its total or average resistance per inch-mil is reduced considerably below (generally to about one-half) the value which it had before treatment.

In constructing, therefore, a very high resistance filament, the necessary total resistance must be secured by making the filaments very long. It is not possible to obtain the requisite durability or life if the diameter of the filament is reduced below a certain value.

The practical problem in the construction of an 8 or 16 c.p. 200-volt lamp is, therefore, to place in a bulb of the usual size a sufficient length of carbon filament to give the desired length. The length of an ordinary 16 c.p. 100-volt carbon filament is about 5in. and it is about 8 or 10 mils in diameter. In an 8 c.p. lamp the length will be somewhat less and the diameter of the carbon will be about 7 mils. The conditions to be fulfilled are determined by the surface area and emissivity of the carbon at the temperature to which it is raised when incandescent, and by the resistivity or specific resistance and length and sectional area of the carbon at the same temperature.

When a treated carbon filament of circular section is at the highest incandescence it can endure in a good vacuum consistently with a duration of an average life of 1,000 hours or so, its temperature is approximately 1,700°C. At this temperature its intrinsic brilliancy is such that it is dissipating about half a watt for every one-thousandth of a square inch of *total* radiating surface. This rate of radiation brings the surface to a state of incandescence at which it emits about one candle light for every 25 ten-thousandths of a square inch of *apparent* surface.*

Thus, for instance, if the total surface of a round carbon filament is 160 square inch mils (1 square inch mil = ·001 of a square inch), the *apparent* surface which radiates in any one direction is about 50 square inch mils. The rate of dissipation of energy is then 80 watts and the candle-power about 20.

Accordingly, the construction of a high-voltage carbon filament lamp of normal candle-power, and contained in a bulb of the usual size, depends very much upon contrivances for getting the requisite length of carbon filament into the bulb. Since primary or untreated carbon has a higher specific resistance than treated carbon, some makers advocate the use of an untreated carbon for use in high-voltage lamps. It is, however, recognised that the softer carbon disintegrates or volatilises faster than the treated carbon; hence, an untreated filament generally deteriorates faster than a treated filament.

If a well-treated filament is used for making a 16 c.p. 230-volt lamp, about 14in. of carbon filament has to be got into the bulb. Lamp makers achieve this in various ways. In some cases they simply place two double-loop filaments side by side, but connected in series in one bulb. In other cases they crinkle or corrugate the single-length filament (*see* Fig. 54). In order to prevent the long carbon loop from touching the glass, or to keep the various turns of the loops from contact

---

* The apparent surface of a circular sectioned filament is equal to a rectangular area having the same length as the filament, and a width equal to its diameter.

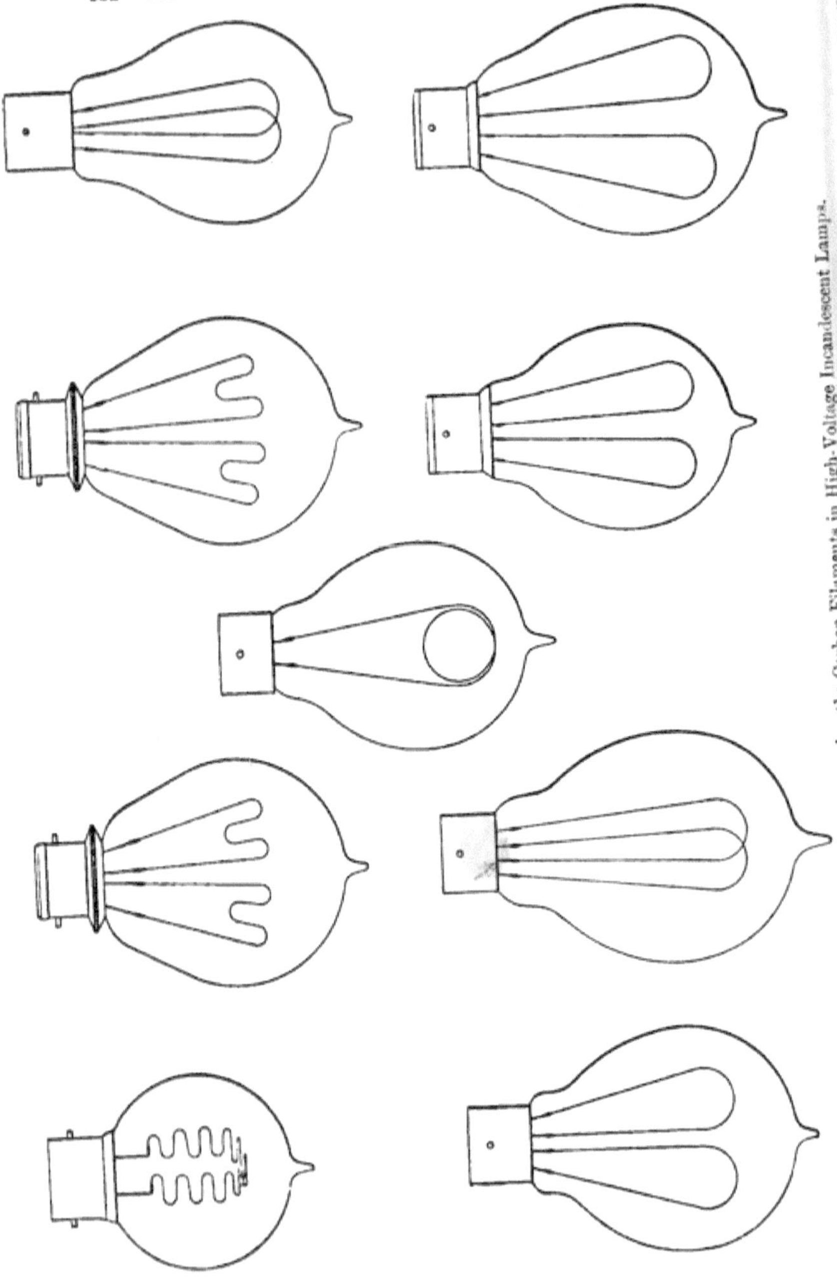

132   ELECTRIC LAMPS AND ELECTRIC LIGHTING.

FIG. 54.—Various Modes of arranging the Carbon Filaments in High-Voltage Incandescent Lamps.

and so from short-circuiting, it is generally found essential to hold the carbon loops in position by stay-pins of platinum wire let into the glass bulb. In any case the carbon filament must be sufficiently stiff when hot, or appropriately held, to prevent it from bending over and touching the glass. If it does touch the glass will be cracked and the vacuum spoilt. High-voltage lamps are now made for 5, 8, 16 c.p. and upwards, up to working voltages of 250 c p.

In the low candle-power high-voltage lamps it is, however, necessary to make the filament so thin that it has not a very satisfactory life unless used with a perfectly constant service pressure.

A few words may next be said on the subject of the "ageing" of carbon filament lamps.

It is a familiar fact that, after a certain use, carbon lamps deteriorate in candle-power. This is due to two causes. The carbon, as explained, is scattered or volatilised and deposited on the glass, and this deposit renders it more opaque to light. Then, also, changes take place in the filament—either an increase in resistance or an increase in surface—which cause it to be reduced in temperature even when subjected to the same terminal voltage. Generally speaking, it is found that if carbon filament lamps are worked at a constant voltage an initial rise in candle-power takes place when used, and then they gradually decay in candle-power. The initial rise is due to a slight resistance of the filament, which makes the lamp to take more current. In the majority of cases, the subsequent decay in candle-power is due entirely to the blackening of the bulb. If the filament in an over-run and blackened lamp is remounted in a new bulb, it returns practically to the same candle-power for the same current as at the beginning of its life. Hence the "useful life" of a lamp is limited by the rate of decay, and a lamp which has deteriorated 20 per cent. in candle-power should be reckoned as worked out.

This rate of decay depends greatly upon the ease with which the carbon surface is disintegrated, hence the softer carbon produced by the simple carbonisation of an organic material is less useful than "deposited carbon" in the manufacture of filaments.

Some tests have shown that, on the average, "unflashed" or "untreated" 16 c.p. filaments lose 42 per cent. in candle-power and 35 per cent. efficiency, or 120 candles per kilowatt, in the course of 600 hours' use.

A comparison between a treated cellulose filament and an untreated bamboo filament 16 c.p. lamp, both taking 3·20 watts per candle when new, showed that there was a progressive fall in candle-power and rise in watts per candle as follows, when measured at regular intervals during the life:—

Table showing Typical Decay of Treated and Untreated Carbon Lamp Filaments with Use.

Hours' run.	Candle-power.		Watts per candle-power.	
	Treated cellulose.	Untreated bamboo.	Treated cellulose.	Untreated bamboo.
0	16·0	16·0	3·16	3·20
100	15·8	14·1	3·26	3·50
200	15·86	12·9	3·13	3·80
300	15·68	11·8	3·37	4·08
400	15·41	11·0	3·53	4·32
500	15·17	10·4	3·51	4·53
600	14·96	9·9	3·54	4·75
700	14·74	9·6	3·74	4·90

To meet the demand for a high-efficiency lamp—that is, a lamp taking less than 3 watts per candle, manufacturers now generally issue lamps of various efficiencies, as, for example, lamp filaments marked for voltages at which they will give light at the rate of 2·5, 3·0, 3·5 and 4·0 watts per candle. Lamps working at less than 3·0 watts per candle are usually called high-efficiency lamps. Any lamp, however, can become a high-efficiency lamp by being used at a higher pressure than its marked volts. Such use, however, is accompanied by a decreased average life.

The following table (from the *Electrical World*, Vol. XXX., 1897) shows the results of some tests on typical 3·1 watt lamps in producing abbreviated life and increased efficiency when the lamps were run at voltages above the normal.

Tests on 100-volt Carbon Filament Lamps when run at Higher Voltages, the Lamps having normally an efficiency of 3·1 watts per Candle.

Working voltage.	Average life.	Candle-power.	Efficiency.
100	1,000	100	3·1
101	818	106	3·0
102	681	112	2·9
103	662	118	2·8
104	452	125	2·7
105	374	132	2·6
106	310	140	2·5

For the sake of ease of comparison the normal candle-power was taken at 100, and the normal life as 1,000 hours, when worked at the marked voltage of 100, and the other values reduced in proportion.

The two qualities of life, or duration and efficiency or watts per candle, are, therefore, closely connected, and the last cannot be changed without affecting the first. It is obvious that this is simply due to the greater rate of volatilisation of the carbon at the higher temperature.

To obtain high efficiency, that is, small watts-per-candle, the incandescent material must be used at a very high temperature, and it is obvious that the limit is imposed, in the case of carbon, by the fact that carbon is one of those substances, like iodine and camphor, which begin to pass into a state of vapour at temperatures lower than their boiling points. Of late years attention has, therefore, been directed to the use of other materials than carbon for the manufacture of the incandescing filament or conductor.

It has been known for a long time that refractory oxides, such as lime, magnesia and other analogous substances, which are practically non-conductors when cold, become changed into fairly good conductors when heated to a high temperature.

A small cylinder of hard lime or compressed magnesia is a very fair non-conductor when cold, but if heated in the oxyhydrogen flame or in an electric arc it becomes changed, so that it can pass current under the operation of an electromotive force of about 400 or 500 volts. Dr. Nernst has lately called attention again to this fact, and, by making small rods of mixtures of refractory oxides, such as the oxides of thorium, zinconium, cerium, yttrium and others, has prepared materials which can become sufficiently conductive after being heated in a flame or by the radiation from an incandescent platinum wire to pass current under electromotive forces of 100 or 200 volts.

Lamps of this description require, therefore, a preliminary heating in order to start them into action. Moreover, since the resistance of the incandescing material decreases rapidly as the temperature rises, the oxide rod has to be joined in series with a metallic resistance wire in order that the lamp may be worked in parallel with others on constant pressure circuits.

Owing to the fact that refractory oxides of the earthy metals can endure an exceedingly high temperature without volatilisation, and are not attacked by oxygen at a high temperature, it is not necessary to enclose the oxide rod in an exhausted vessel, and also it is possible to secure for a lamp of this description a very high efficiency. Experience is, however, as yet (July, 1899), wanting to show what would be the durability or rate of radiative decay of such oxide lamps.

Many more or less successful attempts have been made to find an efficient substitute for the ordinary carbon filament in an incandescent lamp:—

Dr. C. Auer von Welsbach has proposed to employ a filament of osmium or of platinum on which is placed a deposit of thoria.

Langhans has incorporated silicon with carbon, and prepared a carbide of silicon filament.

Maxim has prepared a very dense and adamantine form of carbon, and innumerable patents have been taken out for incorporating with or depositing on the carbon filament some substances more refractory than carbon and less volatile at high temperatures. Up to the present, however, no really successful substitute has been found for a well-manufactured carbon filament lamp as a means of producing light by incandescence.

LECTURE III.

THE Forms of Electric Discharge: Brush, Glow, Spark, Arc.—Vacuum Tubes.—Sparking Distance.—The Electric Arc.—The Optical Projection of the Arc.—The Arc a Flexible Conductor.—The High Temperature of the Arc.—Non-Arcing Metals.—Lightning Protectors. —The Distribution of Light from the Arc.—Continuous and Alternating Current Arcs.—Voltage Required to Produce an Arc.—The Physical Actions in the Arc.—The Changes in the Carbons.—The Distribution of a Potential in the Arc.—The Unilateral Conductivity of the Arc.—The Temperature of the Crater.—Comparison with Solar Temperature.—The "Watts per Candle" of the Sun.—Intrinsic Brightness and Dissipative Power of Heated Surfaces.—Comparison of Glow Lamp, Arc Lamp, and Sun, in respect of Brightness and Radiation.—Arc Lamp Mechanism.—Arc Lamp Carbons.—The Hissing of Arc Lamps.—The Application of Arc Lamps.—Inverted Arcs.—Series and Parallel Arc Lighting.—The Enclosed Arc Lamp.

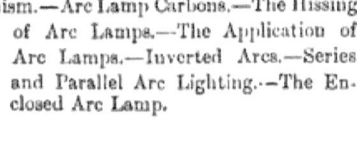

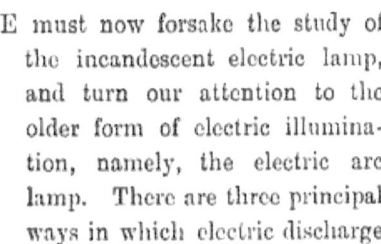

E must now forsake the study of the incandescent electric lamp, and turn our attention to the older form of electric illumination, namely, the electric arc lamp. There are three principal ways in which electric discharge seems to be capable of taking place across a space filled with air or gas. These are: first, by the *glow discharge;* secondly, by an electric spark, or *disruptive discharge;* and thirdly, by the *electric arc.* When two conductors which are at different electric pressures are brought within a certain distance of one another, and when the electric pressure difference between these conductors is gradually raised, one of the above three modes of electric discharge is always established, and takes place in a manner depending upon the gaseous pressure of the air or gas in which the conductors are immersed, upon the electric pressure difference between the conductors (called the electrodes), and upon the strength of the current which the source of the electric pressure can produce. If the two conductors are, say, two brass balls which are connected with the terminals of an ordinary electrical machine, able to make a considerable difference of electric pressure between them, but only capable of generating a very feeble electric current, the following effects are observed when the balls are brought near one another: In the air, at ordinary pressures, the conductors, if viewed in the dark, will be found to be covered with a violet glow of light, and this is more visible if their ends are small—if, for instance, they are the rounded ends of two wires attached to the electrical machine. If one of the conductors has a large surface—if, for instance, it is a brass plate—and the other conductor is not a very good conductor—as, for example, a wooden ball or knob—then in the dark we find a form of glow discharge taking place between the ball and the plate, which is called a *brush discharge*. It is a violet, broom-shaped glow of light, and Sir Charles Wheatstone, on examining it by reflection in a rapidly revolving mirror, showed that it consisted of a series of very rapid electric discharges between the air particles. It is generally accompanied by a slight hissing sound.

This form of brush discharge occurs in nature under some conditions, and is known by the name of St. Elmo's Fire. In the high Alps, when in the neighbourhood of a thunderstorm, travellers have often observed these whizzing brushes of purple light proceeding from their axes and alpenstocks, and from sharp-pointed rocks; and a similar phenomenon has been seen proceeding from the masts and yardarms of ships under certain electrical conditions of the atmosphere. When the knob or ball is positively electrified the electric brush is larger and more brilliant than when the ball is negative. The finest brushes are formed in nitrogen gas. On the other hand, if the conductor is very small, and made of metal, then the glow that is seen on the end of this conductor in the dark is not intermittent, but is associated with a current of air proceeding from the conductor. A blunt metal point attached to an electrical machine can thus be made to blow out a candle flame.

The form and nature of the electric discharge in rarefied gases has been studied by many observers. The phenomena are exceedingly complicated, but generally speaking are somewhat as follows:—Into the extremities of a closed glass tube or bulb are sealed platinum wires, which can be connected to an electrical machine or battery or any source of electric current. If the tube is nearly exhausted, or filled with highly rarefied gas of any kind, and an electric current is passed through it, the following facts can be noticed: A glow of light surrounds each end of the platinum wire, and, if the air pressure is gradually diminished until the air is rarefied to about $\frac{1}{700}$th part of its ordinary pressure, the end of the negative electrode or conductor is seen to be surrounded by a bluish violet light separated from the electrode by a dark space which increases in width as the rarefaction of the air is increased. The positive conductor, or electrode, is also surrounded with a glow of light, frequently of a different colour, and the space in between may be filled up with a glow of light which may or may not be cut up by a series of dark spaces, and is then said to be stratified. This phenomena of the stratified discharge can be shown very beautifully by connecting the two terminals of an induction coil to two platinum wires, which

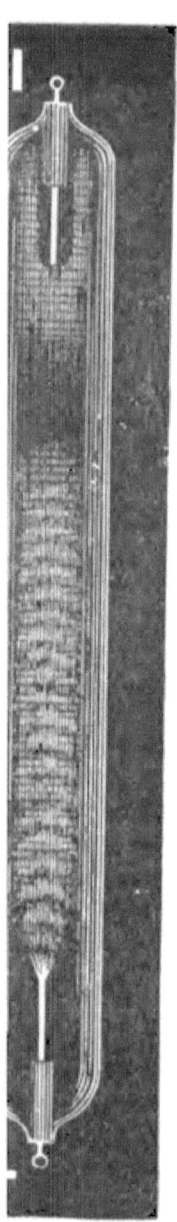

Fig. 55.—Vacuum Tube, showing Stratification.

are sealed into the ends of a long glass tube filled with rarefied carbonic acid (Fig. 55). The tube here used is one of a large collection which belonged to the late Mr. Warren de la Rue, and was presented to this Institution. You will see that the space in between the platinum wires is filled with a light-coloured violet glow, which is stratified or cut up into a number of saucer-shaped layers of light. During this form of discharge the negative electrode, or wire by which the current leaves the tube, is found to become hot, and becomes, in time, raised to a red heat; whilst at the same time, it is more or less disintegrated, depositing a film of metal upon the glass in its neighbourhood.

The glow discharge in gases is a very complex and yet attractive subject, and has engaged the attention of numerous able physicists, such as Faraday, Plücker, De la Rive, Hittorf, Spottiswoode, Moulton, Crookes, and, more lately, J. J. Thompson and many others. It is found that this phenomenon of discharge, both in rarefied gases and in gases at ordinary pressure, can be produced not only by the feeble current of electricity supplied by an ordinary electrical machine or by the intermittent current of an induction coil, but also by the continuous current given by a voltaic battery. When the electric discharge takes place in air at ordinary pressures, and when the supply of current is not very rapid, and especially if one of the conductors is a poor conductor of considerable surface, the discharge takes the form of a series of electric sparks. The form of this spark and the distances over which it can be produced are very much determined by the nature of the surfaces. Rounded metallic knobs give bright snapping sparks, whereas metallic points give much thinner sparks. Besides the form of surface, the interposed gases affect the result. Faraday found that different gases have different restraining powers upon the spark. At the same distance of conductors, hydrogen gas permits a discharge through it more easily—that is, at a lower electric pressure—than air, whereas some other gases, such as hydrochloric acid gas, have peculiar restraining powers. An electric spark is in reality a brief current of electricity taking place between the two conductors, during which time the air or other gas between them is rendered incandescent and therefore luminous, and portions of the material of the conductors between which the discharge is taking place are carried across from one side to the other. For each particular pressure there is a certain *sparking distance*, which, however, is partly determined by the form of the conductors between which the sparks jump. By the term " sparking distance," corresponding to any electric pressure, is meant the distance over which the spark will spring when the difference of electric pressure between the conductors is gradually raised to that value. If we bring two brass balls, connected with the terminals of an electrical machine, within a certain distance of one another, and then darken the room, you will see these two balls surrounded with a glow of light. On bringing them within a certain lesser distance of one another a snapping spark passes, which is repeated at intervals as the machine is worked.

144 ELECTRIC LAMPS AND ELECTRIC LIGHTING.

The first accurate experiments on the sparking distance corresponding to different electric pressures were made by Lord Kelvin. He showed that between slightly curved metal plates it required a pressure of about 5,000 volts to jump across the eighth part of a centimetre, or the twentieth part of an inch. Mr. De la Rue found that between flat plates the difference of pressure required to produce a spark was 8,000 volts when the surfaces were a fifth part of a centimetre, or $\frac{8}{100}$ths of an inch, apart; and 11,300 volts when the surfaces were one-third of a centimetre, or $\frac{2}{15}$ths of an inch, apart. Similar experiments have been made by numerous other physicists with like results.

The following table of results for sparking distances under certain conditions has been calculated from the observations of MM. Bichat and Blondlot. If metal balls $\frac{4}{10}$ths of an inch in diameter (one centimetre) are separated by a distance beginning at $\frac{1}{25}$th of an inch, and increasing by equal amounts—viz., by one millimetre at a time—the electric pressure difference in volts which will just make a spark jump across at the various distances will be as follows (the distances are given in millimetres; one millimetre is nearly $\frac{1}{25}$th of an inch):—

Distances of the balls in millimetres.	Sparking pressure in volts.	Distances of the balls in millimetres.	Sparking pressure in volts.
1	4,765	12	27,024
2	8,140	13	27,765
3	11,307	14	28,359
4	14,119	15	28,949
5	16,664	16	29,363
6	19,210	17	29,837
7	21,823	18	30,133
8	22,792	19	30,547
9	24,153	20	30,932
10	25,071	21	31,198
11	26,255	22	31,494

Hence we see that, between such metallic balls, an electric pressure of 30,000 or 40,000 volts is required to make a spark jump over a distance of about one inch. It has, however, been shown that these sparking distances are greatly affected by the state of polish of the surfaces and by their form, and that sparking is promoted by light of particular kinds falling on the balls. The light from other electric sparks, and also that from burning magnesium wire, has a strong effect in breaking down the insulation of the air and promoting the passage of a spark between polished metal balls across a distance which it would otherwise not pass. Although an electric spark of a few inches in length requires for its production an electric pressure of thousands of volts, it conveys but a small quantity of electricity. It is somewhat surprising to see a torrent of noisy sparks passing between the discharger of an electrical machine, and yet to learn that this represents in quantity, probably, not a thousandth of a coulomb, and that it could hardly effect the chemical decomposition of one drop of water or solution of a metallic salt, although its pressure or potential may be hundreds or thousands of volts. An electric pressure of 100 volts, such as is used for incandescent lamps, will hardly produce a spark over any visible distance; certainly the distance would be less than $\frac{1}{100}$th part of an inch.

These various forms of discharge—the spark, the brush, and the glow discharge—have all been the subject of elaborate investigations. Faraday particularly studied them in the twelfth and thirteenth series of his " Experimental Researches in Electricity." Whatever be the means by which the difference of electric pressure is produced, it is found that, if there exists a sufficient resistance in the circuit limiting the rate at which the supply of electricity can take place, then no other form than the spark, brush, or glow discharge is possible. If, however, we employ a primary or secondary battery—that is, a collection of voltaic cells—for producing the difference of pressure between the two conductors, such an arrangement differs from an ordinary electrical machine only in the fact that it can supply a stronger or more powerful electric current. The voltaic battery, when compared with an electrical machine such as the Wimshurst machine, must be thought of as a very large pump compared with a very small one. Both these pumps can create a difference of level, pumping up water from one level to a cistern or reservoir at a higher level, but the pump of larger capacity differs from the other in that it can supply faster and can keep up the difference of level in spite of out-flow of water. Accordingly, if a large number of voltaic cels are joined together, and the ends of this battery are connected to two brass balls, it is found that, if the battery is one which has a high internal resistance, it will produce a series of small sparks, which jump over continually from one ball to the other, provided the difference of pressure is great enough and the balls are brought near enough together. When the battery is one of low internal resistance, then the spark discharge cannot be maintained; but if the conductors are brought near enough together, and the pressure sufficiently raised, the spark or glow discharge passes spontaneously and immediately into a third form of discharge, which is called the *electric arc*.

It appears that this arc discharge was first observed by Curtet in 1802, only two years after the invention of the battery by Volta; but the first careful study of the phenomena of the electric arc in air and in vacuum was made by Sir Humphry Davy at or about the same time. At that time Sir Humphry Davy was engaged in the remarkable series of electro-chemical discoveries which resulted in the production of metallic potassium and sodium from caustic potash and soda for the first time,

Finding the necessity for a larger battery than he possessed, he laid a request before the managers of the Royal Institution in 1808, asking them to provide the means of constructing a battery of 2,000 cells with which to continue these researches.* Shortly afterwards this battery was provided, and with it Davy not only continued his electro-chemical discoveries, but studied the phenomena of the electric arc. Connecting the ends of this large battery of 2,000 pairs of plates to two pieces of hard charcoal, which had been heated and then plunged into quicksilver to make them better conductors, Davy observed for the first time on a large scale the phenomena of the electric arc. In his "Elements of Chemical Philosophy" (Vol. IV.) he thus describes the first production of this very large 2,000-volt electric arc :—

"The most powerful combination (battery) that exists in which number of alternations (plates) is combined with extent of surface is that constructed by the subscriptions of a few zealous cultivators and patrons of science in the laboratory of the Royal Institution. It consists of two hundred instruments connected together in regular order, each composed of ten double plates arranged in cells of porcelain and containing in each plate 32 sq. in. ; so that the whole number of double plates is 2,000 and the whole surface 128,000 sq. in. This battery, when the cells are filled with 60 parts of water mixed with one part of nitric

---

* Sir Humphry Davy laid a request before the managers of the Royal Institution on July 11, 1808, that they would set on foot a subscription for the purchase of a large galvanic battery. The result of this suggestion was that a galvanic battery of 2,000 pairs of copper and zinc plates was set up in the Royal Institution, and one of the earliest experiments performed with it was the production of the electric arc between carbon poles, on a large scale. It is probable, however, that Davy had produced the light on a small scale some six years before, and, according to Quetelet, Curtet observed the arc between carbon points in 1802. See Dr. Paris' "Life of Sir H. Davy."

acid and one of sulphuric acid, afforded a series of brilliant and impressive effects. When pieces of charcoal about an inch long and one-sixth of an inch in diameter were brought near each other (within the thirtieth or fortieth part of an inch) a bright spark was produced and more than half the volume of the charcoal became ignited to whiteness, and by withdrawing the points from each other a constant discharge took place through the heated air in a space equal at least to 4in., producing a most brilliant ascending arch of light, broad and conical in form in the middle. Fragments of diamond and points of charcoal and plumbago rapidly disappeared and seemed to evaporate in it, but there was no evidence of their having previously undergone fusion."

We may repeat his experiments on a smaller scale. Taking two round rods of hard compressed carbon, these are connected to the two ends of a battery furnishing a pressure of about 50 volts. If these two carbons are brought within a short distance of one another, say $\frac{1}{100}$th of an inch, about the thickness of a thin visiting card, it will be found that this electric pressure of 50 volts is not sufficient to create a spark capable of jumping over this distance. The very thin layer of air existing between the two carbon terminals is sufficient to insulate entirely this pressure. If the two carbons are touched together at the point of contact they immediately become red hot, and then, on separating them again a slight distance, the electric discharge continues to take place between them, and this is called the electric arc. Instead of bringing the carbons in contact we can break down the insulation of the air by passing a small electric spark from one carbon to the other, or even at some distance away. This can be done by taking a discharge from a small electrical machine, and when the experiment is performed before you you will notice that the electric arc discharge follows immediately upon the passage of a spark which is too small to be visible.

A very convenient arrangement for producing and examining an electric arc is one in which one of the carbon rods is fixed and the other is moved by a rack and pinion, so as to be brought in contact with the first, and then moved away as required when the arc is produced between them. This arrangement, one form of which, as devised by Mr. Davenport,* is shown in Fig. 56, is needed in an experimental study of the arc.

Fig. 56.—Hand-Regulated Arc Lamp.

In order to view more comfortably the whole phenomena that are taking place in the interior of this dazzling source of light, we will project an image of the electric arc by means of a lens upon the screen, and examine the nature of the effects taking place in it. The arc discharge is most easily obtained and maintained between conductors which are good but not very good conductors, and which are only

---

* This convenient form of hand-regulated arc lamp was described in *The Electrician*, Vol. XXXII., p. 393.

volatilised at a high temperature. It can easily be maintained between certain refractory metals, or between these metals and graphitic carbon. No substance has been found superior for the purpose of producing an electric arc to the dense variety of carbon, produced either from the graphitic deposit of carbon found in the interior of gas coke ovens, or by the production of a hard variety of carbon from gas coke, lamp-black, or sugar. Twenty-five years ago the carbon rods which were used for this purpose were generally produced by sawing lumps of very hard retort carbon into small rods a quarter of an inch square, but of late years an immense manufacture has sprung up in the manufacture of arc-light carbon rods. In the manufacture of these rods some form of dense and very pure carbon is made into a paste with syrup or coal tar, and from this rods are pressed out which, after being dried and baked at a high temperature, furnish carbon rods of various sizes which are used for the production of the electric arc. These carbon rods are generally sold in lengths, varying in size from a quarter of an inch to an inch in diameter, and from nine to eighteen inches in length. Taking two of these rods, and producing between them an electric arc in a closed lantern, we have now projected upon the screen an enlarged image of the electric arc.

Four things can be noticed in this electric arc as soon as it has been formed by bringing the carbons into contact (*see* Fig. 57). Looking at the optical image of the arc projected upon the screen, we notice that both the ends of the two carbons are brilliantly incandescent, but that, if the arc is formed by a current of electricity flowing always in one direction, called a continuous current arc, then the positive carbon, or the one attached to the positive pole of the battery or dynamo, and marked + in the illustration, is more brilliantly incandescent than the other. This positive pole soon has its end hollowed out into a cup-shaped depression, which is called a *crater*, and this positive or crater carbon, after being in use for a few minutes, will be found to have its extremity converted into graphitic carbon or plumbago, which will mark upon paper like a pencil, in a way that it would not do before being so used. At the

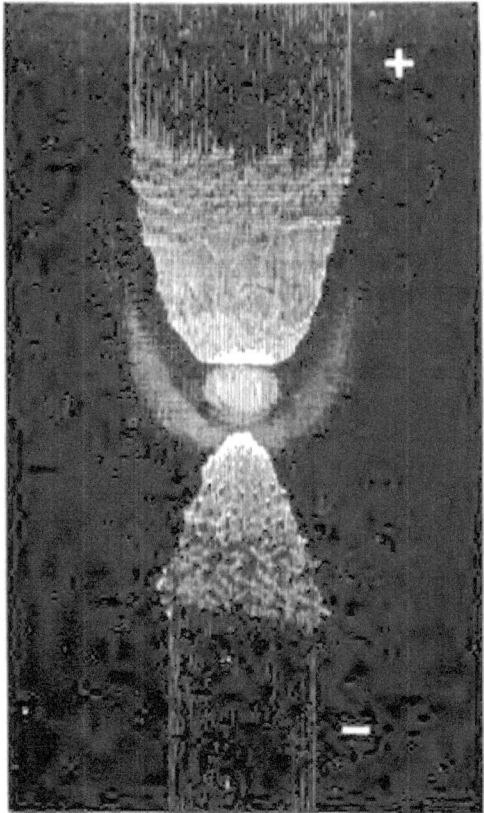

FIG. 57.—The Electric Arc.

same time the negative carbon appears to get more pointed, and becomes surrounded a little below its tip with small globules, which are, in all probability, condensed carbon vapour. If the arc is maintained for some time, it will be found that both these carbons are being worn away; but the positive or crater carbon wears away about twice as fast as the other when the arc is formed in air. In the space between the two carbons we notice a band of brilliant violet light, which is surrounded by a less luminous aureole of a golden colour. In a carefully-regulated, steady electric arc it will be seen that the violet-coloured portion seems to spring from the white-hot crater, and that a well-marked dark space separates this purple part from the golden-coloured, wing-shaped flames which appear to spring from the negative carbon. This intermediate central portion is called the true arc, and hence we have four separate portions of the arc to consider: namely, the crater, the violet-coloured core of the arc, the aureole, and the negative carbon. On looking closely at an arc so projected upon the screen, it will be seen that little bits of carbon are continually becoming detached from one pole, and they immediately fly over to the opposite carbon. It can be shown by several experiments that there is probably a double transport of material going on in the arc. The material of the positive pole is being carried across to the negative side, and the material of the negative pole is being carried across to the positive side.

At this stage it will be interesting to analyse the light of the arc by means of the prism. Placing a vertical slit before the lens which is projecting the image of the arc upon the screen, we cut off all but a linear band of light, the upper part of which proceeds from the crater carbon, the middle part from the violet core of the arc, and the bottom part from the negative carbon. The prism being then placed in front of the lens, we expand this linear image of the slit into a spectrum which is divided longitudinally into three parts. The upper portion, you will observe, is a brilliant continuous spectrum of light, in which all the prismatic rays are present. This spectrum proceeds from the light emitted from the crater of the arc.

The middle portion of the spectrum is much less bright in the orange, red, yellow and green, but, as you will see, is distinguished by two remarkably brilliant violet bands, which are characteristic bands in the spectrum of carbon vapour. The bottom of the spectrum, formed from the negative carbon, is a continuous spectrum similar to the upper one, only less brilliant. On lengthening out the arc by drawing the carbons apart, we find that the two upper spectra move away from one another, the middle spectrum increasing in width. At a certain point the arc breaks down, the middle spectrum with its two violet bands disappearing instantly; but for a short period of time the upper and under spectra linger as the carbons cool, because the carbons continue to glow faintly; and as these carbons cool you observe the two spectra shrinking rapidly from the violet end, until at last only faint traces of the red and green remain, and these finally vanish. But you will see that the lower spectrum, which proceeds from the light of the negative carbon, vanishes first. This only indicates that the crater carbon has a much higher temperature than the negative carbon, and is in accordance with all that we know about the arc.

If we examine an electric arc of this kind, we notice that, as the carbons are gradually drawn apart and the arc lengthened, the aureole of golden vapour increases and forms a sort of large lambent flame, which disappears when the arc is lengthened beyond a certain point. After such a continuous current arc has been extinguished, we can always tell which has been the positive carbon, not only from the crater hollowed in it, but also from the fact that it remains red hot for a longer time than the negative carbon. The exceedingly high temperature of the positive carbon is shown by the fact, above mentioned, that after being used for some time it is converted into graphite at the extremity and will mark paper.

The electrical power taken up in the continuous current electric arc is measured in an exactly similar way to the power taken up and dissipated in the filament of an incandescent lamp —namely, by measuring the current in amperes flowing through the arc, and the difference of electric pressure in volts between the carbons. Multiplying these numbers together, we obtain the value of the power, measured in watts, being absorbed in the arc. By far the best way to denominate electric arcs is by the power in watts absorbed. We may thus speak of a 300-watt arc, or a 500-watt arc, or of a 1,000-watt arc. The length of the arc can be measured most conveniently by projecting the image of the carbons upon the screen. If a sheet of polished metal is then held behind the arc, a faint diffused light will be thrown upon the screen, which will enable us not only to see the arc, but to see also the image of the two carbon rods. By means of a pair of compasses or a foot rule we can measure the length of the magnified image of the arc—that is to say, the distance between the image of the tips of the carbon rods. We can also measure the width of the magnified image of either of the carbon rods. Let us suppose that the rods are each of them half an inch in diameter, and that their image upon the screen is eight inches in diameter; then the arc has been magnified 16 times. If, then, we find that the image of the arc is, say, two inches in length when measured upon the screen, it shows us that the real length of the arc is $\frac{1}{16}$th part of this, namely, one-eighth of an inch, because the length of the arc is magnified in the same ratio as the width of the carbons. This optical method of measuring the length of the arc affords a very easy means of accurately ascertaining the length of the arc without danger or difficulty. Electric arcs may, roughly, be distinguished as long arcs and short arcs, according to the distance between the carbon poles, although the dividing line between the two is not in any way accurately determined.

Before passing on to consider the physical measurements connected with the electric arc, let us notice that the violet core of the arc, which consists of a torrent of incandescent carbon vapour, acts like a perfectly flexible conductor. If I bring a magnet near to such an electric arc, you will notice that the magnet apparently impels the arc sideways, and if it is brought near enough it will actually blow it out. The reason for this is because the arc, as a conductor traversed by a current, exerts an electro-magnetic force upon the magnet, and in like manner the magnet exerts an electro-magnetic force upon the arc. The arc always tends to move across the lines of force of the magnet, and in this way may actually be bent into a bow shape, or be drawn out into a kind of blow-pipe flame.

Brief references must now be made to the generation of light and heat in the arc. Experiment seems to show that in the continuous current arc—that is, the arc produced by a current of electricity always flowing continuously in the same direction—by far the larger proportion, perhaps 85 per cent., of the light thrown out comes from the highly incandescent crater carbon, about 15 per cent. comes from the negative carbon, and very little indeed, certainly not more than 5 per cent., from the true electric arc. The great source of light, therefore, in the electric arc is the intensely luminous crater formed at the end of the positive carbon, which is brought up to an exceedingly high temperature. Various estimates have been made of the temperature of this electric crater, but the physical difficulties connected with these experiments are very great. Rossetti in 1879 stated, as the result of some of his experiments, that the positive carbon had a temperature of $3,200°C.$, and the negative carbon $2,500°C.$ More recently, Violle has given a temperature of $3,600°C.$ as the temperature of the crater carbon, and other observers have given even higher results than these.

This temperature is sufficient not only to melt but actually to boil away even the most refractory metals. We can easily show the high temperature of the crater by boiling in it a piece of copper or silver. If an electric arc is formed with the crater downwards, and a small piece of copper is placed in the crater, and then the image of the arc is projected upon the screen, you will see the little piece of copper boiling violently and sending out a torrent of copper vapour, which, in its incandescent condition, emits a brilliant green light. In the same way, silver, platinum, and even the refractory metal iridium, can be melted and boiled into vapour in the electric arc. It has been already mentioned that there is a transport of material in the arc, material passing over from the positive to the negative pole, and *vice versâ*. It is only natural, therefore, to expect that, since the electric arc is a phenomenon of electric discharge taking place in a highly-conducting vapour, the longest arcs will be capable of being maintained between the most volatile metals, and experience shows this to be the case generally. It has been found that the difference of temperature between the two poles is greater in proportion as they are worse conductors and more easy of disaggregation; also that, to produce an arc of a given length requires a greater current in proportion as the infusibility of the electrode is greater.

A very striking fact was, however, discovered by Wurts, viz., that there are four metals — zinc, cadmium, magnesium and mercury—which are non-arcing metals, so called because it is difficult or impossible to maintain an alternating-current electric arc between the surfaces of these metals which are very near together. This fact is taken advantage of in the production of a very ingenious lightning protector, which may just briefly be described. When overhead circuits, consisting of copper wires carried on insulators on posts, are employed for the distribution of alternating electric currents, such overhead lines are very much exposed to lightning stroke. If they should be struck by lightning, the lightning discharge will generally pass back into the generating station, and do damage to the dynamo machines. In order to prevent this, a lightning protector of the following kind has been designed, especially for use with overhead alternating-current circuits, the object of which is to allow the lightning striking the overhead circuits to pass to earth, but yet at the same time to prevent the current from the dynamo machine following it. On a slab of marble or porcelain are placed an odd number of small zinc cylinders, one inch in diameter and about three inches long. These cylinders are placed close to one another, being separated by a distance only of $\frac{1}{50}$th of an inch. Each cylinder is insulated. The middle cylinder is connected by a wire with a plate sunk in the earth. The two outside cylinders are connected to the two lines which form the overhead circuit. If the lightning strikes the line on either side, the discharge jumps over from cylinder to cylinder until it reaches the middle cylinder, and then it goes to earth by the earth-plate. In so doing it will attempt to start an electric arc between the cylinders; but, as a permanent arc cannot be maintained between zinc surfaces, this arc is almost instantaneously extinguished, and the dynamo machine, therefore, is unable to produce an electric arc across between the two mains. In this way a safe passage for the lightning to earth is provided.

It will, in the next place, be necessary to direct attention to the distribution of light from an electric arc. On examining an electric arc formed by a continuous current, the positive or crater carbon being uppermost, a very cursory examination shows us that the light sent out from such an arc is not uniform in all directions. Very little light is sent upwards. By far the larger portion of the light is sent downwards at an angle of from 30 deg. to 40 deg. below the horizon. By the aid of the explanations given above, it is very easy to see the reason for this. It is really due to the fact that the greater portion of the light is sent out from the bottom surface of the top carbon; hence, when viewed in a horizontal position, but very little of this incandescent crater can be seen. On raising the arc, so as to look up underneath it, the amount of apparent surface of the crater which is seen increases up to a certain point as we elevate the arc. But if the arc is raised beyond a certain amount above the eye, then the bottom or negative carbon begins to stand in the way of the light sent out from the crater, and to diminish the apparent

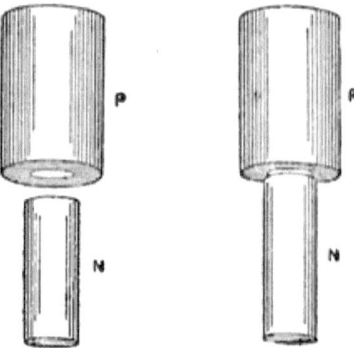

FIG. 58.—Model showing the Relative Areas of Crater seen at different angles of view.

area. By taking two cylinders of pasteboard, N and P, held by threads a little distance from one another (see Fig. 58) and fixing on the underneath end of the upper cylinder a disc of red paper to represent the incandescent crater, we may easily convince ourselves, by looking at this model arc in different directions, that there is a certain position in which the eye can be placed in which it will see the largest apparent area of the crater.

There is, therefore, a position in which the maximum amount of light will be thrown out by a real arc represented by this model. Accordingly, if the intensity of the light given by an electric arc is measured in different directions, say for every ten degrees of inclination above and below the horizontal, we find different photometric values for these rays in these directions. We can represent this different illuminating powers in different directions by a curve (see Fig. 59), which is called the photometric curve of the arc. The magnitude of the radii of this curve, taken in different directions, represent the intensity of the light pro-

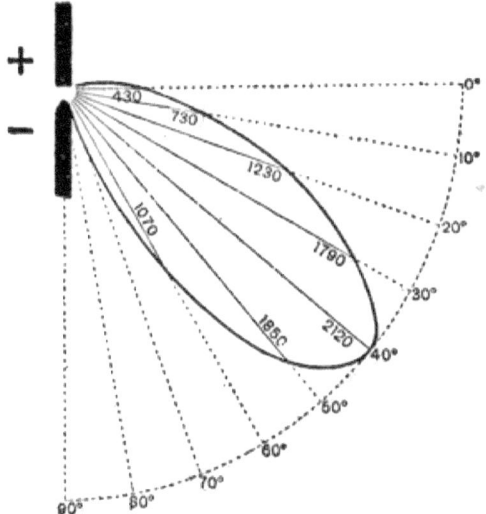

FIG. 59.—Curve showing the Distribution of Light from a Continuous Current Arc. The figures printed on the radial lines represent the candle-power in different directions.

ceeding from the arc in these directions. The exact position of the maximum intensity will depend upon the length of the arc, the size of the crater, and the relative widths of the negative and positive carbons. Accordingly, in street arc lighting advantage is taken of this fact. If the electric arcs are put up on high poles, or street supports, the positive or crater carbon being uppermost, they will throw down their maximum amount of light in a direction inclined at 40 or 50 degrees to the horizon, and illuminate a circular zone around them with a maximum intensity. Just underneath the arc there will be more or less shadow, and in a direction above the horizon very little light will be thrown out.

In order to obtain the best results and the most uniform distribution of light, engineers have found it best to use a very small negative carbon, in order that the negative carbon may stand as little as possible in the way of the light sent out from the crater in a downward direction, and that the arc may, therefore, illuminate the largest possible area underneath it. This unequal distribution of the light from the arc can, to a great extent, be modified by the use of a somewhat peculiar glass globe; but we shall point out in an instant a method by which superior results are gained for internal lighting.

Not only can we form an electric arc by means of a continuous current of electricity—that is to say, one always flowing in the same direction—but we can employ an alternating current of electricity. In such an alternating current there is a very rapid change in the direction of the electric current, the current flowing first in one direction for a short period, then being arrested, and then flowing in the other direction for an equally short period. An alternating current of electricity may therefore be compared to the ebb and flow of water in a tidal river, in which the water flows in one direction for a certain number of hours and is then reversed and flows in the opposite direction for about an equal period. A continuous current of electricity may be likened, on the other hand, to a non-tidal river in which the water flows always in one direction.

In what are called alternating current systems of electric lighting distribution it is very usual to employ an alternating current having a frequency of from 40 to 100 periods per second—that is to say, the electric current changes its direction from 80 to 200 times in one second. We can form an electric arc not only with a continuous but with an alternating current of electricity, and the reason for this is that there is a certain persistence in the electric arc. It is found that, when the arc is made with a continuous current of electricity, the current can be interrupted for a short fraction of a second without putting out the arc. Hence the current of electricity can be reversed in direction without extinguish-

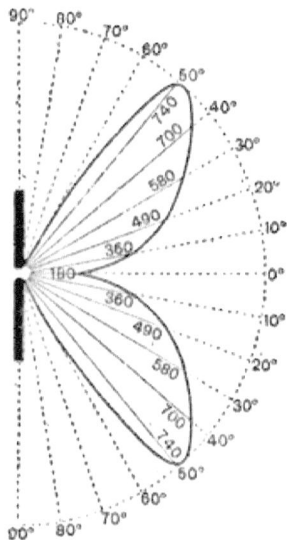

Fig. 60.—Curve Showing the Distribution of Light from an Alternating Current Arc. The figures marked on the the radial lines indicate candle-power in these directions.

ing the arc. If, however, it is reversed very rapidly, then the result will be that, instead of having one positive and one negative pole or carbon, each carbon will be alternately positive and negative, and, instead of having only one crater, we shall have both carbons cratered out and sending out light. An alternating current arc produces, therefore, a very different distribution of light from a continuous current arc. In an alternating current arc the light is not only sent downwards, but it is sent upwards as well, the distribution of light being found to be about equal in directions above and below the horizon. The curve representing the distribution of the light of an alternating arc lamp is shown in Fig. 60, from which it will be seen that the alternating current arc sends its maximum illumination in directions both upwards and downwards equally inclined to the horizon. Objections are sometimes taken to the use of an alternating current arc on the ground that it makes a disagreeable noise or humming. This, however, is much less evident in some arcs than in others, and, by the employment of suitable carbons and a proper frequency of current, can very nearly be abolished. It appears to be proved, however, that an alternating current arc does not yield the same total illumination for a given expenditure of energy in it as does a continuous current arc.

Returning then to the consideration of the continuous-current arc, we may note further physical facts connected with its production between various terminals. If a continuous-current arc is formed between carbon rods, and if the image of the arc is projected on a screen so that its length can be measured, and if, at the same time, the current through the arc and the potential difference (P.D.) of the carbons are ascertained, it is found that there is a definite relation between the length, current, and potential difference which can only be properly expressed by a series of curves.

One important fact in connection with the continuous-current electric arc is that, if the current is increased beyond a certain value, the arc passes into an unstable condition in which it begins to *hiss*. In the case of the open arc this is accompanied by a sudden fall in arc voltage, or P.D., of about 10 volts. On setting out as curves the relation of arc length, current and voltage, it is found that there is a certain minimum voltage (approximately 30 or 35 volts) below which no true arc of any finite length can be formed. This initial voltage has sometimes been called the counter-electromotive force in the arc. There are many reasons, however, which make it impossible to believe that in the electric arc there is a true source of electromotive force working against the external impressed electromotive force.

This initial or non-effective voltage must rather be regarded as one of the factors in the expression for the energy which has to be imparted to the arc to keep up a supply of carbon vapour. The processes which go on during the formation and maintenance of a continuous-current electric arc appear to be something as follows: A large number of facts seem to indicate that electric conduction through vapours or gases is in reality electrolytic in character—that is to say, it depends upon a continual breaking up of molecules of the gas into something smaller, possibly atoms, which are called *ions*. These ions carry electric charges, and when a molecule is broken up into two ions these ions have opposite charges, and move, probably, with different velocities, in opposite directions in the electric field. A similar process has long been recognised as taking place in the case of conduction through certain compound liquids. If this is the case, the column of carbon vapour, which forms the true electric arc, is a region in which resolution of carbon molecules into carbon ions is taking place, and these positive and negative ions are being caused by the electric force in the space to move, probably with unequal speed, in opposite directions.

If an arc, formed with solid carbon rods, is allowed to burn quietly for some time, and if the current is not increased to the hissing point, we have the condition called the normal arc. Under these circumstances, lengthening the arc whilst maintaining the current constant causes the potential difference between the carbons to increase, whilst, if the length is kept constant, increasing the current causes the potential difference between the carbons to decrease. If the current is increased beyond a certain point the arc passes into an unstable condition, and then finally begins to hiss. When this is the case, increasing the current over wide limits hardly at all affects the potential difference of the carbons.

It has been shown by Mrs. Ayrton that the primary cause of hissing is the access of air to the crater. This occurs when the crater becomes so large that it overspreads the end of the positive carbon. The same investigator has also shown that with open arcs there is a sudden decrease of 9 or 10 volts in the potential difference of the carbons at the moment when the arc commences to hiss.

In every arc formed with given carbons and for each arc length there is a maximum current beyond which the arc passes into the hissing state. In the silent state it has been shown by Mrs. Ayrton that the current for a fixed length of arc varies inversely as the excess of the potential difference of the carbons above a certain constant voltage, varying slightly with the length of the arc. The presence of this constant term in the expression for the potential difference of the carbons, in terms of the current and length, has given rise to the theory that the electric arc acts as if it contained a source of electric pressure which operates against the external electromotive force producing the arc. There are various reasons, however, why this position is untenable, and there are also arguments supporting the view that this constant pressure, which must be first applied before any true arc can be formed, represents one factor in the measurement of the physical work that must be done in order to volatilise the carbon in the crater, and to maintain that production of carbon vapour.

Carbon or any other solid material can only be made to pass into the gaseous state by imparting to it a certain quantity of energy usually called the latent heat of vaporisation. In the case of the electric arc the energy so required can only be taken from the current by creating a fall in voltage or potential. Hence, if at the surface of the positive carbon the carbon is being boiled away into vapour, energy absorption, represented by a fall in potential, must occur at that place. Until this carbon vapour is manufactured out of the solid carbon electrode, there is no possibility of the production of a true electric arc, consisting, as this does, in the passage of a current through a column of carbon vapour. It appears evident, therefore, that before any arc of finite length can be produced there must occur a fall in voltage. Since the energy required to volatilise a unit mass of carbon is a constant quantity, and since the quantity of electricity which can be conveyed by a carbon atom, molecule or ion is a constant quantity, it would appear that the absorbed voltage on this counter-electromotive force, as it has been called, must also be a constant quantity. It is this which constitutes the 35 or 40 volts required to begin a carbon electric arc.

In connection with this question we may note some experiments by Victor von Lang. In his experiments he formed electric arcs between poles of various metals, and measured this constant pressure difference below which no arc of measurable length can be maintained. His results are seen in a table on the next page, and it will be at once noticed that the order in which the metals stand as regards the magnitude of this so-called back electromotive force is the order of their infusibility or volatility.

Table of Electric Pressures or Voltages required to begin an Electric Arc between poles of the materials named. (*Victor von Lang.*)

Material used for poles between which the Arc is formed.	Initial voltage for production of an Arc between these poles.
Carbon	35 volts.
Platinum	27 ,,
Iron	25 ,,
Nickel	26 ,,
Copper	23 ,,
Silver	15 ,,
Cadmium	10 ,,

If, then, the temperature of the crater is the temperature of the boiling metal or material beween which the electric arc is formed, we should expect such a crater to have a constant temperature and to emit light of a perfectly constant composition. Reasons have accordingly been given by different physicists for believing that, in the carbon arc, the temperature of the crater of the positive carbon has a perfectly constant value, which is that of the boiling point of carbon.

From the above table of results it will be seen that the most infusible material (carbon) requires the highest voltage to begin a true electric arc, and generally that the order of the initial voltage is the order of infusibility of the materials. It has been shown that the lower the electric conductivity of the material of which the poles are formed, the greater is the difference of temperature between the positive and negative poles when an arc is produced between them. The following, therefore, are the reasons why carbon, in its hard graphitic form, is so peculiarly suitable for use as the poles or electrodes in forming an electric arc :—

1. It is a rather poor conductor of electricity (compared with metals). Hence we obtain a great difference of tem-

perature between the poles; in other words, we gain the advantage of having a highly-heated crater on one pole.

2. It is easily disintegrated. Hence we obtain a longer arc for a given electric pressure difference between the poles.

3. It is very infusible, and its oxide is a gas; and therefore does not form a deposit of oxide in and around the arc.

4. It is an inferior conductor for heat. Hence the high temperature is localized at the ends of the poles.

Just as no material has yet been found which is superior to carbon for the incandescing conductor of glow lamps, so no substance has been yet discovered superior to carbon for use in the production of an electric arc.

Hence we must conclude that the processes at work in the production of an electric arc are somewhat as follows :—In the crater, carbon is being boiled into vapour at a temperature probably about 3,500° and 4,000° centigrade. On looking carefully at a magnified image of the crater projected by a lens on to white paper, we can see an apparent seething and effervescence of the surface of the intensely white hot floor of the crater. This torrent of carbon vapour passing towards the negative pole forms the violet core of the arc, the violet colour being that of the vividly-incandescent carbon vapour, possibly superheated to a higher temperature than the boiling point of carbon by the passage of the electric discharge through it. At the cooler negative pole some of the carbon vapour is condensed; but a portion of the torrent of carbon molecules carried across is deflected back and returns to the positive pole, causing the golden aureole or flame, and creating thus a double carbon current in the arc. The negative carbon gives evidence after use of having been worn away by a kind of sand-blast action, and is so worn away into a series of terraces. The negative carbon also undergoes a curious series of changes, by which a point or knob of carbon is gradually formed on it.

This knob or mushroom end finally breaks off, and is thrown out of the arc. In Fig. 61 are shown a series of diagrams illustrating the gradual changes the negative carbon undergoes.

It was shown by the author in 1890 that, if a third carbon pole is dipped into the continuous current arc, so that its tip is in the core of the arc but not touching either positive or negative carbon, or if the electric arc is projected sideways by a magnet on to the third carbon, then a great difference of electric pressure exists between the positive or crater carbon and this

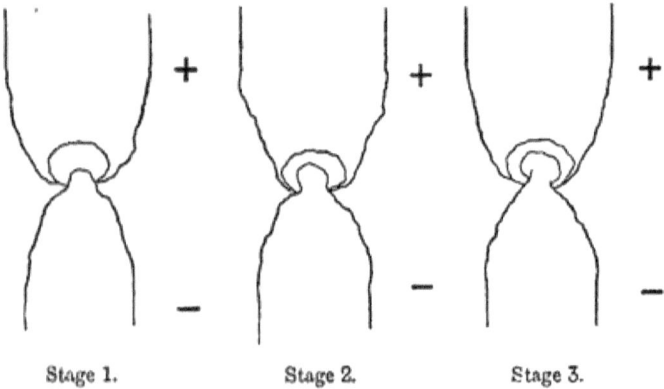

Stage 1.   Stage 2.   Stage 3.

Fig. 61.—Diagrams showing the Changes in Form of the End of the Negative Carbon.

third pole, sufficiently great to illuminate a small incandescent lamp or to ring an electric bell; but that there is a much less difference of electric pressure between the negative carbon and the third pole.

If a continuous current electric arc is formed in the usual way, and if a third insulated carbon, held at right angles to the other two, is placed so that its tip just dips into the arc (*see* Fig. 62), we can show a similar series of experiments to those described in Lecture II. under the head of the "Edison effect." The arc is rather more under control if we cause it to be projected against the third carbon by means of a magnet. We have now formed on the screen an image of the carbon poles and the arc between them in the usual way. Placing a magnet at the back of the arc, the flame of the arc is deflected laterally and is blown against a third insulated carbon held in it. There are three insulated wires attached respectively to the positive and to the negative carbons of the arc and to the third or insulated carbon, the end of which dips into the flame of the arc projected sideways by the magnet. On starting the arc this third carbon is instantly brought down to the same electrical potential as the negative carbon of the arc, and connecting an amperemeter in between the negative carbon

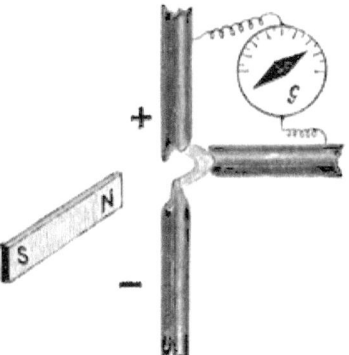

Fig. 62.—Electric Arc projected by magnet against a third carbon, and showing strong electric current flowing through a galvanometer, G, connected between the positive and third carbon.

and the third or insulated carbon, we get, as you see, no indication of a current. If, however, we change the connections and insert the circuit of the galvanometer between the positive carbon of the arc and the middle carbon, we find evidence, by the violent impulse given to the galvanometer, that there is a strong current flowing through it. The direction of this current is equivalent to a flow of negative electricity from the middle carbon through the galvanometer to the positive carbon of the arc. This shows that the fall of potential between the carbons chiefly takes place near the positive carbon. So strong is the current flowing in a circuit connecting the middle carbon with the positive carbon that we can, as you see, ring an electric bell and light a small incandescent lamp when these electric current detectors are placed in connection with the positive and middle carbons.

We also find that the flame-like projection of the arc between the negative carbon possesses a unilateral conductivity. Joining a small secondary battery of fifteen cells in series with the galvanometer, and connecting the two between the middle carbon and the negative carbon of the arc, just as

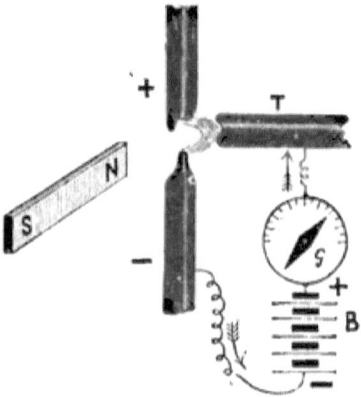

FIG. 63.—Galvanometer G and battery B inserted in series between negative carbon of electric arc and a third carbon to show unilateral conductivity of the arc between the negative and third carbons.

in the analogous experiment with the incandescent lamp, we find we can send negative electricity along the flame of the arc one way but not the other. The secondary battery causes the galvanometer to indicate a current flowing through it when its negative pole is in connection with the negative carbon of the arc (see Fig. 63), but not when its positive pole is in connection with the negative carbon. On examining the third or middle carbon after it has been employed in this way for some time, we find that its extremity is cratered out and converted into graphite, just as if it had been employed as the positive carbon in forming an electric arc.

Much interesting work yet remains to be done in studying the physics of the electric arc, and we cannot say yet that we fully understand the mechanism which underlies this familiar electric phenomenon.

We must not pass on from this part of the subject without making reference to experiments which have been made to determine the intrinsic brightness of the crater of the electric arc. It has been estimated by Mr. A. P. Trotter that each square millimetre of the incandescent crater sends out light equal in brightness to 170 candles. Other observers, such as M. Blondel, have determined a similar value, viz., 160 candles per square millimetre. That is to say, an arc light crater, having an area, say, of 30 square millimetres, would send out light which would produce an illumination or brightness on a white surface placed parallel to it equal to about 5,000 candles placed at the same distance. The intrinsic brilliancy of the arc crater exceeds that of any other source of light with which we are acquainted, except the sun. Certain materials exist for making a comparison between solar temperature and brilliancy, and that of the arc light crater. It is well known that the sun's edge (limb) is much less bright than the centre (40 per cent. less); but, as a first rough approximation, we may take it as equally bright, and assume that the sun radiates light as if it were an incandescent circular disc of 852,900 miles in diameter. The actual area of its apparent surface is then $1\frac{1}{2}$ billion billion square millimetres ($1\cdot 48 \cdot 10^{21}$ sq. mm.). Prof. Young, in America, has made an estimate of the illuminating power of the sun, and his conclusion is that it is equal to 1,575 billion billion candles ($1575 \cdot 10^{24}$ c.p.), correction being made for terrestrial atmospheric absorption.

Hence the mean intrinsic brilliancy of the sun's apparent surface is about 1,000 candles per square millimetre.

The rate at which energy is being sent out from the sun's surface has been estimated from data given by Herschell and Pouillet; and more recently Forbes and Langley have corrected these data, so that from them Lord Kelvin estimates the solar radiation as equal to 133,000 horse-power per square metre. This corresponds very nearly to 100 watts per square millimetre. Accordingly every square millimetre of solar surface is sending out energy as radiation at a rate equal to 100 watts, and producing a candle-power of 1,000 candles. The sun is, therefore, working at an "efficiency," as a glow-lamp maker would say, of $\frac{1}{10}$th of a watt per candle. If we take the mean value of the estimates that have been made for the intrinsic brilliancy of the crater of the electric arc as 160 candles per square millimetre, and the rate of energy radiation as 30 watts per square millimetre, which is probably not far from the truth,* we see that the solar surface radiation is 100 watts, and the solar intrinsic brilliancy 1,000 candles per square millimetre; whilst the electric arc light crater has a radiation of about 30 watts and a brilliancy of 160 candles per square millimetre. The brilliancy of the sun is, therefore, according to these figures, six times that of the electric arc crater, and its rate of surface radiation about three times. Compare this again with the incandescent carbon of the glow lamp. A 16-c.p. glow lamp with circular filament has a carbon filament five inches long and $\frac{1}{100}$th of an inch in diameter. The area of

---

*A series of experiments made in the author's laboratory seemed to show that in various sized arcs, using from 300 to 1,000 watts, the crater area was always of such size that in every case about 27 to 30 watts was dissipated per square millimetre of crater surface. This is on the assumption that the whole work done in the arc is in the first place expended in the crater, and that the rest of the heating of the carbon is due to secondary effects.

its apparent surface is, then, about 30 square millimetres, its total surface is very nearly 100 square millimetres, and it will radiate energy at a rate equal to about 50 watts when giving a light of 16 candle power. Hence, per square millimetre of apparent surface, its intrinsic brilliancy is about half a candle, and its rate of radiation half a watt. Comparing the radiation power and intrinsic brilliancy of the three illuminants—sun, electric arc crater, and glow lamp filament—per square millimetre of surface, we have a rough comparison as follows:—

	Rate of radiation of energy per square millimetre of surface.	Intrinsic brilliancy or candle-power per square millimetre.
Sun	100 watts.	1,000 candle-power.
Electric Arc Crater	30 ,,	160 ,,
Glow Lamp Filament	$\frac{1}{2}$,,	$\frac{1}{2}$,,

The law connecting radiation with temperature at these high temperatures is not yet definitely known, but the above figures warrant the conclusion that the solar temperature, though much higher, is not enormously higher than that of the crater of the electric arc. The temperature of the sun has sometimes been assumed to be millions of degrees Centigrade. All such guesses must be exceedingly wide of the mark. The solar temperature, at least at the surface, is probably much nearer 6,000° or 7,000°C.

It is necessary to note the proper manner of comparing various illuminating agents in respect of their specific energy radiation and light-producing powers. If we consider a straight cylindrical carbon filament which is, say, 5in long and $\frac{1}{100}$th of an inch in diameter, this filament will have a *total* surface of nearly $\frac{15}{100}$ths of a square inch, but its apparent or projected surface will only have an area of $\frac{1}{20}$th of a square inch. Let this filament, when incandescent, give a candle-power of 16 candles measured in a direction at right angles to itself, and absorb a total power of 48 watts. If we take as our unit of area $\frac{1}{1000}$th of a square inch, called an inch-mil, it is clear that the energy waste is at the rate of 48 watts for 160 units of surface, or 1 watt per $3\frac{1}{3}$ inch-mils. The candle-power given by the circular section filament is, however, just the same as would be given out by a flat strip of carbon, kept at the same temperature, the length of which was the same and the width of which was equal to the diameter of the round filament. Hence the *brightness* of the surface, or candle-power given out perpendicularly by each unit of surface of the round filament, is obtained by dividing the whole candle-power—viz., 16—by the area of the projected surface of the circular filament, viz., $\frac{1}{20}$th of a square inch. If, therefore, the unit of surface is the $\frac{1}{1000}$th of an inch, the brightness of the surface is $\frac{16}{50}$ths of a candle per inch-mil, the energy waste is at the rate of $\frac{48}{50}$ths of a watt per inch-mil of surface.

In the usual way of reckoning what is called the "efficiency" of the glow lamp it is usual to divide the whole power taken up in the filament (in this case 48 watts) by the resultant or observed candle-power in one direction (in this case 16 c.-p.), and to obtain a quotient (in this case 3) of watts per candle-power. If we are, however, defining the rate at which energy is dissipated per square unit of area of the incandescent surface, and the *brightness* or normal candle-power per square unit of the surface, we see that the ratio of the numbers representing brightness and power radiated per unit of area is not the same as the ratio of the numbers representing the observed candle-power and the total power dissipated when we are dealing with glow lamp filaments. If we call the total power in watts radiated per square unit of area from the incandescent surface the *dissipating power* of that surface, we see that the ratio of dissipating power to brightness is about one watt per candle for a carbon filament when worked under the conditions of temperature at which its "efficiency" would generally be called three watts per candle-power. Hence it happens that in the above table, comparing solar, arc, and glow lamp radiation, the brightness of the carbon filament is given as half a candle per millimetre, and the dissipating power as half a watt per millimetre. It is necessary to distinguish between the ratio of dissipating power to brightness and total radiation of energy to observed candle-power. We shall call the former the watts per normal candle-power and the latter the watts per observed candle-power.

We cannot assume that the law which experiment shows connects together the candle-power and the rate of energy waste in watts in a glow-lamp filament is followed at temperatures much higher than that at which the incandescent filament is usually worked. The temperature of the carbon filament when being used at about 3 watts per candle-power is probably not far from 1,500°C. or 1,700°C. Up to that point the candle-power appears to increase nearly as the cube of the power in watts taken up. We cannot, however, heat the carbon to a temperature higher than that which corresponds to one watt per observed candle-power without rapidly volatilising it in vacuo. In the electric arc crater the temperature appears to correspond to about one-fifth of a watt per normal candle-power; and this temperature M. Violle has stated to be about 3,600°C., or about double that of the incandescent carbon of the glow lamp at normal temperature. These estimates of arc crater temperature, however, probably err on the side of being too low, and the true crater temperature may be even higher than the above value. Stefan, in 1879, suggested that the temperature of a heated black body and its rate of radiation of energy were connected as follows: The fourth power of the absolute temperature of the body varies as the rate of radiation of energy from a unit of the surface of the body. In other words, if the rate of radiation of energy is increased 16 times the temperature of the surface would be doubled, if 81 times trebled, and so on. It has been asserted by several experimentalists that this law is inapplicable to high temperatures, but if we assume for the moment that it can in any degree apply to the range of temperature extending from the glow lamp to the solar surface temperature, then, since the dissipating powers of the surfaces of the glow lamp filament, arc crater, and sun, are, from the table above, seen to be nearly in the ratio of $1 : 50 : 200$, the absolute temperatures should be in the ratio of $1 : 2\frac{2}{3} : 3\frac{3}{4}$ roughly, and this would indicate the solar surface temperature as being not much greater than 7,000°C.

We must now briefly consider the practical side of electric arc lighting. We shall make no attempt to enter into details as to the countless forms of arc lamp mechanism, but in very general terms, explain the nature of that mechanism. From what has been said above, it will be seen that, in order to maintain an electric arc between carbon points, some form of apparatus has to be devised which will automatically perform the three following operations:—First, bring the carbon rods or points together when no current is passed between them; second, as soon as a current passes, separate them to a determined distance, or "strike the arc," as it is termed; and, third, will bring the carbons together slightly if the current decreases, or separate them slightly if it increases. This last action is called "feeding the arc." In a good arc lamp mechanism the "feed" is very smooth and uniform, and is marked by an entire absence of all jerky movements. All early attempts at the invention of arc lamp mechanism comprised the use of clockwork, and this involved many objections. Modern arc lamp mechanism almost entirely depends upon the employment of electromagnets, which act as the source of power to move the carbons. Without attempting any extended description of even a portion of the devices employed, a very general idea may be obtained by considering the simplest form of shunt and series coil lamps. One of the carbon rods, say the upper one, is fixed to an iron rod or core I (*see* Fig. 64), which has one end inserted just inside the hollow of a bobbin wound over with thick insulated wire, Se, and the other end inserted just inside a bobbin, Sh, wound over with fine wire. The

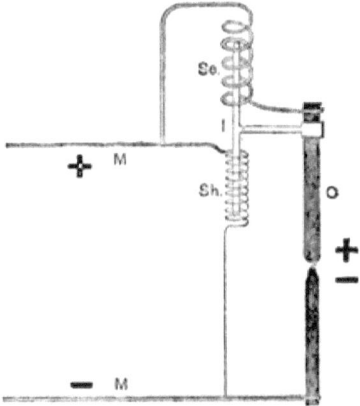

Fig. 64.—Diagram of Shunt and Series Coil Arc Lamp Mechanism.
M, M, are the two Electric Supply Mains; Se, the Series Coil; Sh, the Shunt Coil; I, the Iron Core, and C, the Carbons.

former bobbin is inserted in the line, bringing the current to the arc, and is called the *series* coil. The fine wire coil has its wire joined across between the two carbons, and is called the *shunt* coil. The arrangement will be understood by reference to Fig. 64.

When the carbons are placed in contact, and a current is passed through them, this current passes through the series coil on its way to the carbons. It therefore energises that coil, and causes it to attract or pull in the iron core. It will be understood by reference to the figure that this action immediately separates the carbons a little way. At the same time the shunt coil is traversed by a current, the action of which is to attract its half of the iron core, and therefore to exert an influence drawing the carbons together. The carbons therefore become separated by a small amount, thus striking the arc, and the amount by which they become so separated is determined by the relative magnetic power of the two coils. If the current becomes weakened by the carbons burning away, then the action of the shunt coil preponderates, and draws the carbons together again, thus shortening the arc and increasing the current. When the current increases in strength the series coil exerts the most powerful action of the two, and separates the carbons again by a small amount. It will be seen that the iron core is kept floating between two balanced attractions, one of which tends to draw the carbons attached to it together, the other to separate them. By this simple means, modified in practice by various details of construction, an arc lamp can be constructed which is self-feeding and which sets itself in action by the simple process of switching it on to the circuit.

There are an immense variety of arc lamps depending upon this principle, in all of which various means are taken to make the feed of the lamp properly gradual. If the feed of the lamp is not sufficiently regular it shows itself by rapid changes in the appearance of the arc. As the arc lengthens the violet light of the true arc preponderates, and the lamp burns with a much more bluish colour. If the carbons then come together suddenly the lamp is very liable to hiss and to burn for some time with a more reddish light until the carbons are separated. In lamps with imperfect feeding mechanism these changes give rise to irregularities in the light which are very annoying. In a good arc lamp the feeding mechanism should work with such perfect regularity that the movement of the carbons can hardly be discerned with a magnifying glass. The best manner of examining the feed of an arc lamp is to project, by means of a lens, an image of the arc lamp upon the screen, and then to examine this optical image. The motion produced by the feeding mechanism is then magnified in the same ratio as the dimensions of the arc, and any irregularity of motion is very easily detected.

One word ought to be said with regard to the quality of carbons used in arc lighting. A vast amount of ingenuity and capital has been expended in investigations intended to discover the best processes for the manufacture of arc light carbons. These carbons are now made as above mentioned by forming into a paste some very pure form of carbon by means of a fluid, such as tar or syrup, capable of being carbonised. This paste is squeezed out into rods by powerful hydraulic presses, and these rods are then dried and baked. Carbons are classified as cored or non-cored carbons. In the cored carbons the centre of the carbon rod is perforated by a longitudinal hole, thus forming a carbon tube, and this carbon tube is filled up with a less dense form of carbon. These cored carbons are generally used for the positive carbon of the arc, and the object of making a softer central portion to the carbon is to determine the position of the crater and compel it to keep a central position.

As already explained, in operating continuous current arc lamps the positive or crater carbon is generally placed uppermost. Under these circumstances there is a cone of shadow directly under the lamp which is due to the negative carbon. Mr. Crompton has found it to be a great advantage to use very much smaller negative carbons than is usually done, and in his arc lamp he uses the smallest negative carbon which can be employed practically. On the other hand, a too small negative carbon is liable to become heated over a greater portion of its length, and therefore a practical limit is placed to the use of very small negative carbons.

The quality of carbons used and their form have a great effect upon the performance of an arc made with them. Violent movements of the arc and periods of hissing or humming are very objectionable in the case of electric arcs used for illuminating purposes. These undesirable actions are under control to a large extent, and depend much upon the quality of carbons used. In the arc lamps used for illuminating purposes it is generally usual to employ a positive or upper carbon about 14 or 15 millimetres in diameter, and a negative or lower carbon about 10 or 11 millimetres in diameter. If the arc is a 10-ampere arc the current density is then about 50 amperes per square inch of carbon section. If the current density is too great the crater will be too large and will extend over the edge of the positive carbon. Access of air will then cause the arc to begin hissing. At the moment when hissing begins the difference of pressure between the carbons falls and the current increases. The true arc then loses its transparency and becomes transformed into an incandescent mist which hides the crater. When hissing ceases the arc again becomes violet and transparent, and the surface of the crater appears to be covered with black specks which gradually disappear. Whilst hissing lasts the brightness of the crater is diminished and the violet colour of the arc becomes replaced by a very characteristic blue green, which seems to point to a lowering of the temperature.

It is usual to use an upper or crater carbon which is *cored*, that is to say, has a longitudinal hole through it filled up with a softer carbon. This has the effect of maintaining the crater in a central position and preventing the wandering motion of the crater, which is in many ways objectionable.

Numerous experiments have been tried, by using carbons cored with various materials or saturated with various salts, to improve the quality of the light. On the whole, however, the most satisfactory result is obtained by the use of the purest carbons.

In an ordinary 10-ampere arc lamp, as used for public lighting, the carbons employed each burn away at the rate of about one inch per hour. This disappearance of the carbon is largely caused by actual combustion of the carbon by the air. A lamp of this type used all night, every night of the year from dusk to dawn, consumes about 600ft. of carbons per annum. If the positive carbon is an 18 millimetre carbon and the negative a 12 millimetre the consumption of each size of carbon is about 70ft. per thousand hours of burning, including wastage.

At present prices the carbons for 1,000 hours of burning cost, wholesale, about 15s. Hence, if an arc lamp using such carbons is burnt from dusk to midnight every night the cost of carbons is about £1. 10s. per annum. These carbons are sold in lengths of 15in. or 18in., and at regular intervals an arc lamp requires to be trimmed or re-carboned. In the case of public arc lighting one man and one boy can trim about 60 lamps per day unless the weather is very bad. The cost of the labour per lamp is about £2 per annum. Hence, for labour of trimming and carbons the cost of maintenance of a public arc lamp is about £3. 10s. to £4 per annum.

If the lamp is a 10-ampere lamp it will use 500 watts per hour or one Board of Trade unit in every two hours of burning. If arc lamps are lighted at dusk and burn until midnight the annual hours of usage are 1,800 to 2,000. A 500-watt arc will, therefore, use 900 to 1,000 Board of Trade units per annum, and if this energy is supplied, say, at 3d.

per unit, the annual cost of electric energy for a 500-watt arc lamp burnt from dusk to midnight will be £11 to £12, and, including carbons, labour for trimming and repairs, will cost £15 to £16 per annum to maintain. Street arc lamps are always mounted on high standards, 20ft. to 25ft. above the roadway. These are placed from 50 to 100 yards apart. The following table gives some data as to the distance, hours used, height above roadway and annual price charged for arc lamps in various towns :—

Town.	Distance apart of arc lamps in yards.	Total number of hours' use in year.	Price charged per arc per annum. £ s. d.	Height of arc above roadway in feet.	Current in amperes.
Blackpool	50	2,000	—	25	10
Brighton	70	3,864	—	—	—
Bristol	50	—	—	22	10
Cardiff	55-70	3,824	18 0 0	—	—
Chelmsford	80	3,687	22 10 0	—	—
Derby	45-90	3,500	—	—	—
Hanley	50-140	3,900	16 0 0	—	—
Kingston	110	3,400	20 0 0	—	—
Newport	70	2,787	25 0 0	—	—
Portsmouth	90-100	1,800	18 0 0	25	12

In making calculations as to the cost of running arc lamps it is useful to have at hand the total hours of burning per annum when the lamps are lighted and extinguished at certain hours as follows :—

Hours of lighting.	Annual hours of burning.
Sunset to dawn	4,374 hours
Sunset to 11 p.m.	1,821 ,,
Sunset to midnight	2,186 ,,
Midnight to sunrise	2,188 ,,
11 p.m. to sunrise	2,553 ,,

In the employment of arc lamps for illuminating purposes, owing to the fact that the light is radiated from an exceedingly small area (in a 300-watt lamp, about $\frac{1}{84}$th part of a square inch), the light sent out by the lamp throws very sharply defined shadows if the lamp is used without a shade. Hence it is usual to enclose the lamp in a large semi-opaque globe. These globes cut off from 40 to 60 per cent. of the light in various directions; but, by presenting a larger light-giving area they cast less well-marked shadows and are less dazzling to the eye. On the other hand, this wasteful absorption of the light causes the lamp to be less efficient as an illuminating agent. In order to meet this difficulty the device is frequently adopted of employing inverted arc lamps. In this arrangement the arc lamp is hung from the

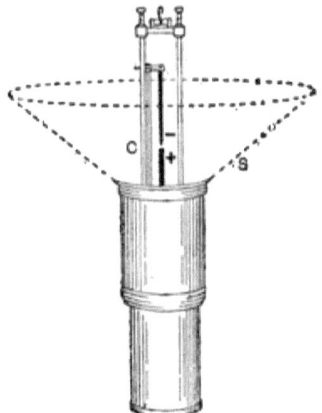

FIG. 65.—Inverted Arc Lamp and Shade for Workshop and Factory Lighting. The dotted line shows the place of the Conical Reflector.

ceiling or roof of the building to be illuminated with the negative carbons uppermost (see Fig. 65), and the positive or crater carbon is placed at the bottom of the pair. The lamp is surrounded with a conical shade, the purpose of which is to throw the light up, and in addition sometimes a white umbrella-like shade is suspended over the lamp, from which the light is again reflected downwards. A room illuminated by these inverted arc lamps is therefore flooded with light, none of the rays of which come directly from the arc, but only after reflection from the hood or the ceiling. A very pleasing, diffused light is thus produced, and inverted arc lighting has been tried, in many cases with great success in workshops, dye works and factories. You are able to judge of the effect of it yourselves if this theatre is illuminated for a few minutes by means of a couple of inverted arc lamps which have been lent to me by Messrs. Parsons and Co. If a workshop is illuminated by arc lamps in the ordinary manner, even although they are protected by semi-opaque globes, the light casts sharp shadows around and under the tools, and the workmen are therefore sometimes unable to get the light necessary for their work; but the inverted arc lighting does away with this difficulty, and in places so illuminated there are practically no shadows at all.

The necessity of re-carbonising arc lamps used for public lighting every day has turned the stream of invention of late years in the direction of improvements to obviate this necessity, with the result that we now possess very efficient enclosed arc lamps which will burn with one pair of carbons for 150 or 200 hours. It has already been explained that the consumption of the carbon rods is largely due to the actual combustion by the air of the incandescent tip of the carbons. The problem is to prevent this combustion. At first sight nothing would seem more simple, the apparently obvious solution of the difficulty being to enclose the arc in an air-tight glass globe. This device was tried very early in the history of electric lighting. The result of using a completely air-tight globe is that the contained oxygen is removed by combustion with the carbon, and some of the carbon vapour forming the arc diffuses through the vessel and is deposited on the cool sides of the containing vessel as a black deposit. The transparency is thereby affected. This always ensues if the containing vessel is perfectly

enclosed. It was, however, shown by Marks that if the
vessel was not completely closed it was possible to admit

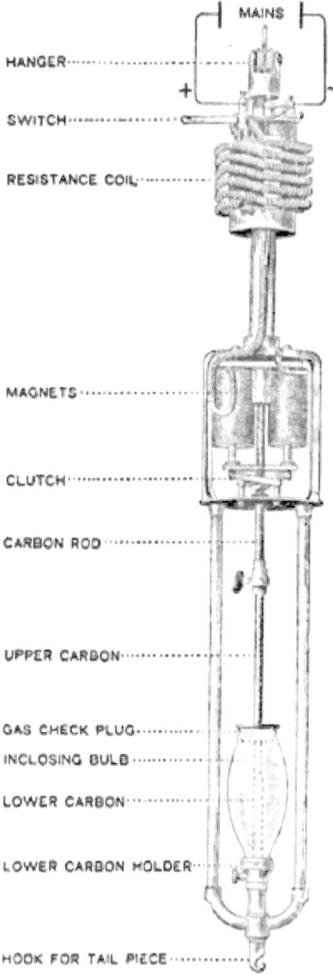

FIG. 65A.—The Marks High Potential Enclosed Arc Lamp.

air in such a limited manner that it destroyed by combustion
the carbon vapour escaping from the arc, but was not
admitted freely enough to cause very wasteful oxidation of

the carbon rods themselves. An arc lamp so enclosed in a glass receptacle, with a limited access of air to the arc, is called an *enclosed arc lamp*. In Fig. 65A is shown an enclosed arc lamp The top of the enclosing bulb is closed by a gas check plug or plate (*see* Fig. 65B), which

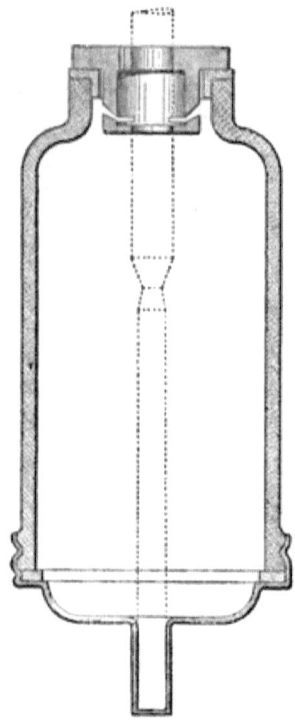

Fig. 65B.—Enclosure with Gas-Check Plug.

admits, through a small hole, a limited supply of air to the arc, sufficient to consume the volatilised carbon, yet not sufficient to cause very great consumption of the carbon rods themselves.

The marked peculiarity of the enclosed arc lamp is that the carbons do not burn to a crater on the positive, and a sharp tip or mushroom on the negative, but they maintain a nearly flat surface (see Fig. 65c). This feature affects the distribution of light. The illuminating curve of the enclosed arc is not characterised by such a strongly marked maximum value, as is that of the open arc. We have already seen that the open arc sends out rays in a direction enclosed at about 60° below the horizon, which have generally three times or more the intensity of the horizontal ray. Also there is a very great difference between the

Open Arc : 6 amperes, 45 volts.   Enclosed Arc : 6 ampere-, 45 volts.
FIG. 65c.

mean spherical candle power of an open arc lamp and its maximum candle power. It is for this reason, amongst others, that all such terms as "an arc lamp of 2,000 candle power," are perfectly meaningless. In the case of the enclosed arc lamps, there is not, however, such a great difference between the mean spherical and maximum illuminating power. The enclosed arc emits its rays of greatest illuminating power more in a horizontal direction.

Summarising the advantages of the enclosed arc, they are as follows:—

(1.) The cost of carbons per hour is about $\tfrac{1}{20}$th of that for the open arc.

(2.) The cost of attendance is also far less.

(3.) There is a more uniform distribution of light from the arc.

Against these advantages must be set:—

(1.) The cost of renewals of the inner glass vessel which occasionally breaks.

(2.) The lower resultant candle power per watt, due to the absorption of light by the inner glass vessel and outer globe, which is also necessary to diffuse the light.

It is found by experience that if the inner and outer globes are both of opalescent glass, that the total loss by absorption of light is more than 50 per cent. The best results are obtained by using an inner opalescent vessel and a clear glass outer globe.

The enclosed arc lamps as now made burn without attention for 150 or 200 hours on 100-volt circuits. Small arc lamps are made taking 3, 4 and 6 amperes, and are very convenient for interior illumination. For indoor use it is essential to enclose the inner vessel in a glass spherical globe, which may be of opalescent glass, to soften and diffuse the light, but for outdoor use it is best to employ an opalescent inner vessel, and a clear outer globe. Enclosed arc lamps are now also made to burn on 200-volt circuits.

Enclosed arc lamps are constructed for use with alternating as well as continuous currents. Generally speaking, however, the alternating arc lamp has not the same efficiency as the continuous-current arc lamp—that is to say, for the same mean spherical candle power the alternating-current arc lamp will require a greater expenditure on it of electric power. Great attention has been paid of late years to the phenomena of the alternating-current arc. If an arc is formed between carbon poles with an electric current which is not continuous, but which changes its direction or alternates 100 or 200 times a second, then an arc is formed which is called an alternating-current arc. The electric

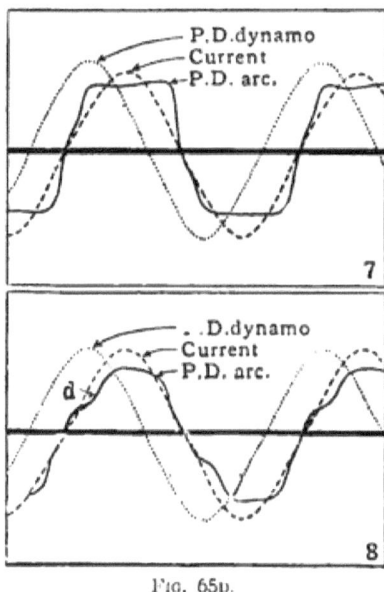

Fig. 65d.

arc has a certain persistence, so that the current can be interrupted for a very small fraction of a second, without extinguishing the arc. If, then, the current alternates, each carbon in turn becomes the crater carbon, and during the short period of time it is the positive or crater carbon, it emits more light than when it is the negative carbon. If an alternating arc is examined through slits in a revolving disc which is driven in step with the alternations of the arc, it is possible to examine the various phases of light-radiating power passed through by each carbon. It is then seen that there is a periodicity or fluctuation in the light-giving power of each carbon. This variation in the light given out by each carbon follows the variation of current through the arc. The arc has, however, a peculiar power of re-acting upon the impressed potential difference or external electromotive force. If the machine (called an alternator) which supplies the current is such that its electromotive force varies periodically in the manner represented by the ordinates of the dotted curve shown in Fig. 65D, then the potential difference and the current through the arc may, and generally will, vary in a totally different manner, as shown by the broken and firm line curves in the same diagrams. It has been shown that under some conditions the arc formed between a metal pole and a carbon pole furnishes a current only in one direction, although the impressed electromotive force is an alternating arc.*

We cannot, in conclusion, do more than make brief allusion to the manner in which arc lamps are employed in practical work. One method is by the employment of arc lamps arranged in series. In this arrangement the arc lamps to the number of 30 or 50 are placed on one circuit, so that the same electric current, say one of ten amperes, flows through each lamp in turn. As each lamp requires to have about 50 volts on the terminals in order to make it work properly, it will be seen that the arrangement necessitates the use of a dynamo machine giving a very high electric pressure—from 1,500

---

* The reader, desirous of more advanced information, as to the alternating-current arc, may be referred to the following original Papers: "Experiments on Alternating-Current Arcs by the aid of Oscillographs," by W. Duddell, E. W. Marchant, *Journal* of the Institution of Electrical Engineers, Vol. XXVIII., 1899, or *The Electrician*, Vol. XLII., p. 857. Also "An Analytical Study of the Alternating-Current Arc," by J. A Fleming and J. E. Petavel; *Phil. Mag.*, April, 1896, p. 315.

to 3,000 volts. Series arc lighting is employed to a considerable extent in street lighting, for the reason that the conductors conveying the current can be, comparatively speaking, small and inexpensive. Arc lamps may, however, be arranged *in parallel* instead of in series. In this case two or more lamps are placed together across the mains between which a constant electric pressure is being maintained. Suppose, for instance, in any district an electric lighting supply company has the mains laid down for giving a supply at 110 volts for incandescent lighting, then two arc lamps can be placed across these mains in series, a small resistance being added in order to control the flow of current through the lamps. At the same time any number of these pairs can be arrranged across between the mains like the rungs of a ladder. In some cases five to ten arc lamps are worked in series across mains having a difference of potential between them of from 250 to 500 volts, a number of these series of five or ten being strung across in parallel between two supply mains. An objection which has always been urged against series arc lighting is that it necessitates the employment of high pressure currents, which are, of course, relatively much more dangerous than low pressure currents. In series arc lighting each lamp has, moreover, to be provided with an apparatus called an automatic cut-out, so that if the lamp becomes extinguished by the carbons being separated too far or burning out, the cut-out closes the circuit, and does not permit the current which is supplying the other lamps in the series to be interrupted. In the case of alternating-current arc lamps, they may either be run in series on a high tension alternating-current circuit, as is done in the electric lighting of Rome, or each lamp may be provided with a small transformer, so that current can be taken from high tension mains and reduced down to a pressure at which it is convenient to work the arc lamps. We shall in the next Lecture proceed to explain more in detail the nature of this pressure-reducing device.

LECTURE IV.

THE Generation and Distribution of Electric Current.—The Magnetic Action of an Electric Current.—The Magnetic Field of a Spiral Current.—The Induction of Electric Currents.—The Peculiar Magnetic Property of Iron.—Iron and Air Magnetic Circuits.—The typical cases of an Iron Circuit with and without Air Gaps.—The prototypical forms of Dynamo and Transformer.—The Transformation of Electrical Energy.—Hydraulic Illustrations.—The Mechanical Analogue of a Transformer.—The Mode of Construction of an Alternate Current Transformer.—The Fundamental Principle of all Dynamo Electric Machines.—Alternating and Direct Current Dynamos. —Alternating Current Systems of Electrical Distribution.—Description of the Electric Lighting Station in Rome.—The Tivoli-Rome Electric Transmission.—Views of the Tivoli Station.—Continuous Current Systems.—The Three-Wire System.—Description of St. Pancras Vestry Electric Lighting Station.—Liverpool, Glasgow, and Brussels Electric Lighting Stations.—Direct-driven Dynamos. —Alternating and Continuous Current Systems Contrasted.—Secondary Batteries. — House Meters. — Maximum Demand Meters.—Conclusion.

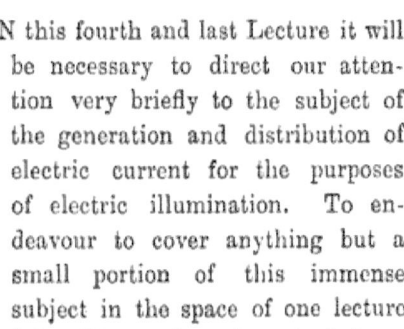

N this fourth and last Lecture it will be necessary to direct our attention very briefly to the subject of the generation and distribution of electric current for the purposes of electric illumination. To endeavour to cover anything but a small portion of this immense subject in the space of one lecture would be to confuse and bewilder rather than to inform. It will, therefore, be necessary to limit our discussion to certain narrow lines, and not to attempt, in any way, to give even an outline of the whole of the methods which are at the present moment employed for this purpose. Avoiding technicalities, however, the endeavour will be made to enable the reader to grasp certain broad general principles which underlie the construction of all the appliances used for the generation and distribution of electric current for industrial purposes, and then we shall proceed to describe generally the arrangements employed in the electric lighting of a modern city. We must return, in the first place, to facts touched upon in the First Lecture, and examine more in detail the relation of magnetism and electric currents. In so doing it may be an advantage, for the sake of brevity and clearness, to depart a little from the usual phraseology. We must also trace our way in the light of a series of preliminary experiments.

It has already been pointed out that a wire, conveying what we call an electric current, exerts a magnetic influence, and at the same time becomes heated. In fact, the phrase, "an electric current," is merely a comprehensive expression used to denote all the effects of heat and magnetism that are known to exist in and around a wire which we speak of as being traversed by an electric current. If we take a conducting wire —say a copper wire—and pass through it a strong electric current, we find that such a wire, when dipped into iron filings, takes them up, the filings clinging to the wire forming a bunch around it. This fact was discovered about 1820 by the French experimentalist Arago. We can more carefully and deliberately explore the nature of this magnetic action of a wire conveying a current in the following way : Passing the copper wire through a hole in a glass plate, or card, so that the wire stands vertically to the plate, we then sprinkle the glass plate uniformly with steel filings, and, by causing a powerful current to pass through the copper wire, and by gently tapping the glass plate, the filings arrange themselves in a series of concentric circular lines round the wire (*see* Fig. 66). The glass plate being placed in this vertical lantern, the image of these circular lines of filings is projected upon the screen. Taking a very small magnetised compass needle, and holding it in any position near the glass plate, it at once becomes evident that the little magnetised needle always places itself in the same direction as the circular lines mapped out by the iron filings. This fact may be

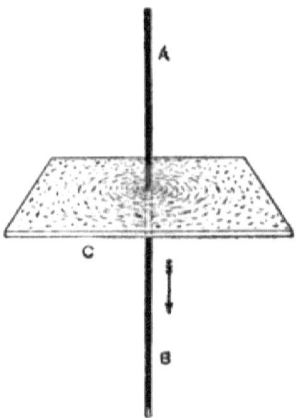

FIG. 66.—Curves formed by sprinkling Iron Filings on a Card **C**, perforated by a Wire conveying an Electric Current, thus showing the lines of Circular Magnetism round the Conductor **A B**.

viewed, in the following light: The circular magnetisation of the space round the wire is a manifestation of a part of the properties of the electric current, and may properly be called the **magnetic part of the current**, and is called the magnetic flux surrounding the conductor. In fact, the magnetic flux is a part of that which we call the electric current. The direction of this magnetic flux round a conductor may be ascertained at any place by the use of iron filings as above described.

o 2

A very little examination of the distribution of the iron filings shows us that they lie closer together at points near to the wire than at points far away.

It is important to examine another instance of the same kind. Here is a large bobbin of wire, the wire being covered over with cotton and wound on a paper tube. Into the interior of this bobbin is placed a slip of glass on which steel filings have been sprinkled. On passing an electric current through the wire and tapping the glass, we find that the steel filings arrange themselves inside the bobbin in a series of nearly parallel straight lines (*see* Fig. 67), and that, in this particular case, it is, therefore, evident that the electric current,

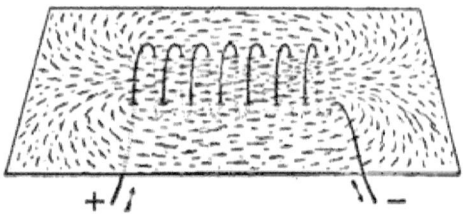

FIG. 67.—Curves formed by Iron Filings sprinkled on a Card, through holes in which a spiral wire is laced, showing the lines of Magnetic Flux or Induction linked with a Spiral Electric Current.

flowing through the wire in a series of nearly circular turns, is accompanied by a distribution of magnetism, or a magnetic flux, in a direction *along* the axis of the bobbin. By exploring the space outside the bobbin with a magnetic compass needle it is very easy to show that the magnetic flux which flows in one direction in the interior of the bobbin flows back along the outside of the bobbin in an opposite direction, and so completes what is called a *magnetic circuit*. Whatever particular form we give to the conductor conveying the electric current, we always find that there is a magnetic flux in the space round it along a magnetic circuit which is linked with the conducting circuit. This co-linking of electric conducting circuits and magnetic circuits is the fundamental fact of electro-magnetism.

So far the cases we have been examining have been such that the magnetic flux exists in a circuit which is wholly occupied by air as a medium. If, instead of permitting the magnetic flux to take place in air, it takes place in iron wholly or partly, then it is found that the same electric current flowing in the wire would produce a much more powerful magnetic flux in the iron circuit than it does in the air circuit.

It is necessary to examine this point with some care, and to do this we must define a little more accurately the mode of measuring the magnetic quantities concerned. If a wire covered with silk or cotton is wound any number of times round a bobbin of any kind, and if a current of a certain strength, measured in amperes, is allowed to flow through that wire, the product of the number of turns of the wire and the strength of the current measured in amperes which is flowing round the bobbin wire is an important quantity, which is called the *ampere-turns* of that bobbin. The total number of ampere-turns on the bobbin, divided by the length of the bobbin, gives us the ampere-turns per unit of length of the coil. We have already seen that in the interior of such a bobbin of insulated wire when traversed by an electric current, there is a magnetic flux—or magnetic induction, as it is called—the direction of this field being in the direction of the axis of the bobbin. Before we can show how this induction is measured it will be necessary to direct attention to the fundamental discovery in connection with this subject.

Imagine that in the interior of the bobbin employed a moment or two ago in our experiments another bobbin of wire is inserted, so as to occupy a portion of the space in the interior. Let this second bobbin be called, for the sake of distinction, the secondary conducting circuit, whilst the first is called the primary circuit. Faraday discovered that, as long as the electric current flowing in the primary circuit remains constant or unaltered, no effect will be produced by it upon the secondary bobbin. The aperture, or central hollow, of this secondary bobbin remains traversed by part or all of the magnetic induction, or, as we have called it, the magnetic flux of the first bobbin, but otherwise it is not affected. If the electric current traversing the primary bobbin is either stopped or reversed in direction, or is altered in strength, the immediate result is to produce a current of electricity circulating in the secondary bobbin; which current is, however, only a very brief or transitory current, and does not continue for any length of time. It is, as it were, a sort of wave of electricity passing through the secondary circuit. By appropriate instrumental means we can measure the quantity of electricity which is in this way set in motion. We can also measure, in the unit called an ohm, as explained in the First Lecture, the electrical resistance of this secondary circuit. Faraday showed that the value of the magnetic induction which is traversing the secondary bobbin could be exactly measured by the product of the resistance of this secondary circuit and the whole of the quantity of electricity set flowing in it when the primary current was suddenly arrested. Hence the use of a secondary circuit of this kind affords us the means of exploring the magnetic condition of the space in the interior of such a primary bobbin.

In Fig. 68 is shown a pair of spiral conducting circuits, P and S, which consist of spirals of wire wound through suitable holes in a card. If a current of electricity is passed through the circuit P, and if iron filings are sprinkled on the card, we find that the lines of magnetic induction, due to the primary circuit, are marked out. Some of these lines will be seen to pass through, or be linked with, the secondary circuit S.

It may be well to state at this point that the quantity of electricity which flows past any point on a conducting electric circuit in one second, when it is being traversed by a current of one ampere, is called *one coulomb*. We are able to measure quantity of electricity by means of an instrument called a ballistic galvanometer, and, if we are provided with such an appliance, it is a comparatively simple matter to determine the whole quantity of electricity which flows through a circuit when any change takes place in the magnetic current or induction linked with it.

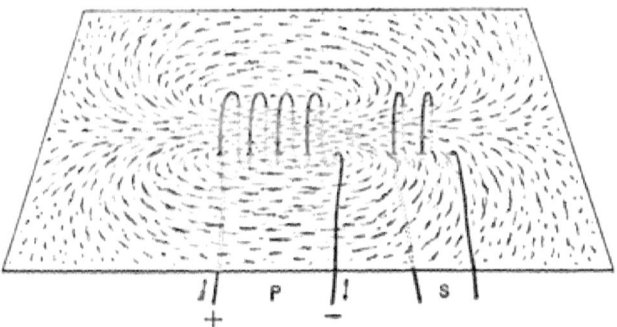

Fig. 62.—Curves delineated by Iron Filings on a Card, showing the lines of Magnetism of a Spiral Primary Circuit, P, passing through and linked with a Secondary Circuit, S.

Returning, then, to the case of our primary and secondary bobbin: let us, for the sake of simplicity, assume that the secondary bobbin is wound closely on the outside of the primary bobbin, and that the primary bobbin has a certain given number of turns, and is traversed by a current of a certain strength measured in amperes. When a primary current of, say, ten amperes is passing through the primary bobbin, we have a magnetic flux of a certain strength, or a magnetic induction taking place in the direction of the axis of that primary bobbin, part or all of which magnetic induction or flux perforates or passes through the secondary circuit. If the secondary circuit is connected to a ballistic galvanometer, and the primary current is suddenly stopped, the galvanometer will give an indication that a brief wave of electricity has passed through the secondary circuit, and this is called a secondary current. The product of the quantity of electricity in this secondary current and the resistance of the secondary circuit is, under these circumstances, a measure of the amount of the total magnetic induction, or the total magnetic flux due to the primary current in the primary bobbin which is flowing through or linked with the secondary circuit. In the above case the magnetic circuit consists of air—that is to say, the magnetic flux takes place in a circuit which is merely the air space in and around the primary bobbin.

Supposing that we insert in the interior of the primary bobbin a thick iron rod wholly filling up the interior space, and either straight or bent round so as to form a closed ring. Experiment shows that under these circumstances the state of affairs is greatly altered. On passing through the primary bobbin the same current of, say, ten amperes, and then stopping that current suddenly, we should find that the electric impulse produced in the secondary circuit is immensely increased, and that the quantity of electricity circulated in the secondary circuit would then be a thousand or more times greater—multiplied probably 50 times if the iron bar were a short straight rod, and multiplied 2,000 times if the iron was bent round in the form of a ring. This fact shows us that a given number of ampere-turns in the primary bobbin produces an electric impulse, or electromotive force, in the secondary circuit which is dependent upon the nature of the material filling the space in the interior of, and outside of, the primary bobbin.

These facts can all be summed up in the following statements: If there be two circuits of insulated wire, which are called respectively the primary and secondary circuits, and which are placed near to one another or wound over one another, then the flow of a continuous current of electricity through the primary circuit produces no effect whatever upon the secondary circuit, so long as that primary current remains unaltered in strength. If the primary current is changed in strength, either being increased or diminished, reversed or reduced to nothing, this change in the number of ampere-turns in the primary bobbin gives rise to an electromotive force, producing an electric current in the secondary circuit, and this electromotive force depends essentially upon the *rate of change* of the current strength, or the rapidity of the change which is made in the strength of the primary current, and, therefore, in the rate of change of strength of the magnetic flux or magnetic induction which is produced by the primary current, and also passes through the secondary circuit.

Following a conception of Faraday's, it is usual to speak of the direction of the magnetic flux which surrounds the electric current as the direction of the magnetic *lines of force* due to the primary current, and these lines of force can, as we have already explained, have their directions made manifest by the employment of steel or iron filings. The peculiar property which iron possesses, when used as a magnetic circuit, is that it permits the production through it of a given magnetic flux by the employment of a far less number of ampere-turns per unit of length than does any non-magnetic material.

We may examine the behaviour of iron in this respect most easily by returning to the fundamental experiment by which it was discovered. Let us take a wooden ring having a circular section, and wind over it insulated copper wire to form a primary circuit, and over that again another insulated secondary circuit. Through the primary circuit let an electric current be passed, producing, therefore, a certain magnetising force, measured by the number of ampere-turns per unit of length of that bobbin. From what has been above explained it will be seen that this primary current gives rise to a magnetic induction or magnetic flux which passes round the ring, and some or all of which traverses the secondary circuit. If the primary current is stopped, the arrest of the magnetic flux flowing through the secondary circuit creates in it an electromotive force which causes a brief flow of electricity to take place through that circuit, if it is complete, and that electromotive force at any instant is

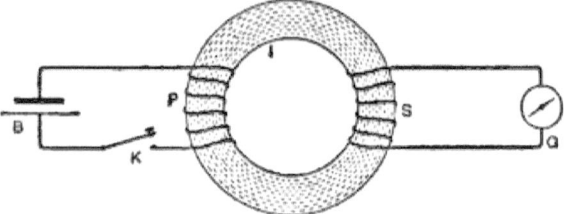

FIG. 69.—Iron Ring I, having wound on it two Circuits,—a Primary, P, connected to a battery, B, through a switch or key, K, and a Secondary Circuit, S, connected to a galvanometer, G. When the Primary Current is stopped or started it creates an Induced Current in S.

measured by the rate at which the magnetic flux or magnetic induction passing through the secondary circuit is being changed. Supposing, however, that these circuits, instead of being wound on a wooden ring, are wound on an iron ring of exactly the same size (see Fig. 69), and that we pass through the primary circuit a current of the same magnitude as before. The arrest or stoppage of this current would now be found to produce a vastly greater electromotive force in the secondary circuit, and the explanation of this fact is that the primary ampere-turns are now able to produce a much greater magnetic flux through the secondary circuit, and that, therefore, the arrest of the primary current creates a greater total change in the magnetic flux flowing through the secondary circuit. The nature of the material filling the space, either partly or wholly, in and around the primary and secondary circuits, has a very important influence on the inductive effect of the primary circuit upon the secondary.

It is not unusual to state that the primary circuit exerts a *magneto-motive force* on the magnetic circuit. In the above simple instance of the iron ring, we see that we have two electric circuits and one magnetic circuit, which are linked together like the links of a chain, and that between the three circuits there is a fixed relation of action which is as follows: The electromotive force acting in the primary circuit produces in the primary circuit an electric current. This electric current, flowing a certain number of times round the primary circuit, produces a magneto-motive force of a given magnitude, which creates a magnetic flux, or, as it is called, a magnetic induction in the magnetic circuit. The magnetic circuit, in turn, exercises an influence upon the secondary circuit, but only when the magnetic induction or magnetic flux is being changed. In the secondary circuit an electromotive force is set up whenever the magnetic flux is being altered in magnitude, and the electromotive force acting to produce an electric current in the secondary circuit is measured by the rate at which the magnetic flux or magnetic induction throughout is being changed.

We have, therefore, the following operations related to one another in these three circuits, the primary circuit, the secondary circuit, and the magnetic circuit. The primary electromotive force acting in the primary circuit produces a primary current; the primary current acting on the magnetic circuit produces a magneto-motive force; the magneto-motive force produces a magnetic flux or magnetic induction in the magnetic circuit; the magnetic induction, *when changing in amount*, produces a secondary electromotive force in the secondary circuit, and therefore a secondary current. In order to distinguish that quality of iron and air in virtue of which a given magneto motive force produces a much greater magnetic current or magnetic induction in the iron than it does in the air, we say that the iron possesses a less magnetic *reluctivity* than the air. The magnetic *reluctivity* of iron may be as much as from one to two thousand times less than that of air.

Supposing, in the next place, we make a cut across each side of our iron ring (as seen in Fig. 70), separating the iron ring into two portions by a narrow air-gap across each

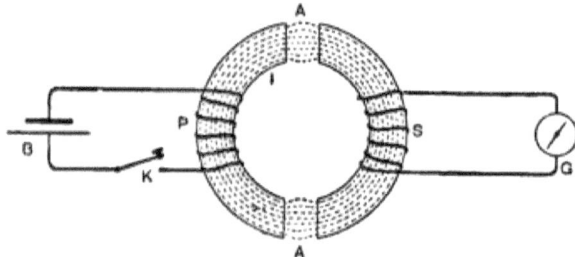

FIG. 70.—Iron ring I, as in Fig. 69, but having a pair of air-gaps made by cutting the ring in two at the points A A.

side. It would then be found that a given electric current flowing in the primary circuit would, when arrested, produce a much smaller secondary electromotive force in the secondary circuit than it does when the iron ring is uncut. The reason for this is that the air-gap on both sides has increased the magnetic resistance of the magnetic circuit; and, accordingly although the magnetic flux or induction is able to traverse the air-gaps, the magneto-motive force of the primary circuit is now not able to produce so much magnetic flux or magnetic induction in the magnetic circuit; and hence, when the primary current is arrested, the change in the magnetic induction or magnetic flux flowing through the secondary circuit is not so great as before.

ELECTRIC DISTRIBUTION.   205

The two cases we have just examined—namely, the iron ring complete, with a primary and a secondary circuit wound over it, and an iron ring with two air-gaps in it—that is to say, two half-iron rings, on one of which is wound a primary circuit, and on the other a secondary circuit—constitute two typical models which will enable us to explain the action of appliances which are called dynamo-electric machines and transformers.

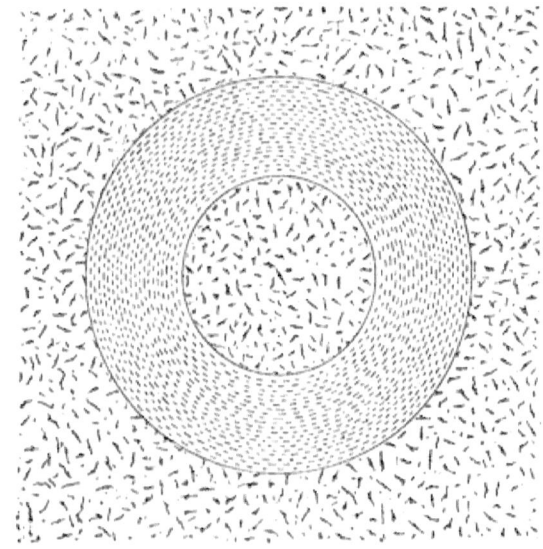

FIG. 71.—Iron Ring magnetised in lines round the ring, but producing no external magnetic field or influence.

Attention should, at this point, be directed to four diagrams, which are reproductions from photographs of experiments easily made.

In Fig. 71 is shown an iron ring, circular in form. Let a spiral of insulated wire be wound on this ring and an electric current be passed through it. This current will magnetise the ring along lines shown as dotted lines. If iron filings are sprinkled over a card held over the ring they will not arrange themselves in any particular form in the space outside the ring, thus showing that the magnetism is wholly confined to the ring. Next, let a narrow cut be made in the ring (*see* Fig. 72). At once we find, by applying the iron filings test, that there is now a development of magnetism in the space outside the ring, and that there are lines of magnetic induction or flux passing across from one side of the air-gap to the other, producing a powerful magnetic

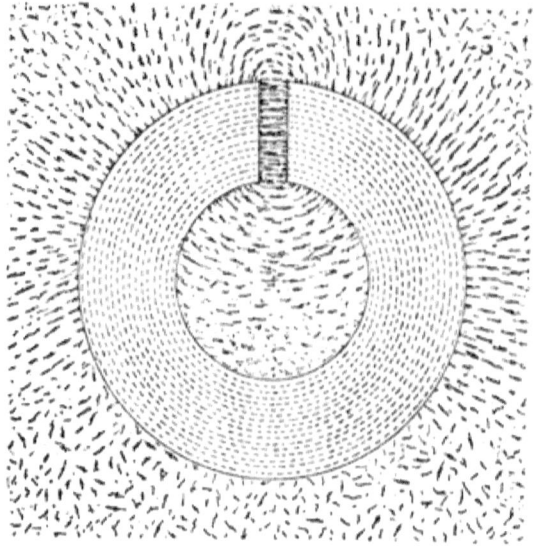

Fig. 72.—Magnetised Iron Ring with cut or air-gap. The ring is magnetised along the dotted lines. The external magnetic field or induction is delineated by iron filings.

field in the gap space. By making wider gaps, as shown in Figs. 73 and 74, the varying forms of the external fields can be detected by the use of the sprinkled filings.

We have already seen that, in order to produce an induced electromotive force, and therefore an electric current, in the secondary circuit, it is necessary to produce a change in the magnetic flux or magnetic induction passing through that

ELECTRIC DISTRIBUTION. 207

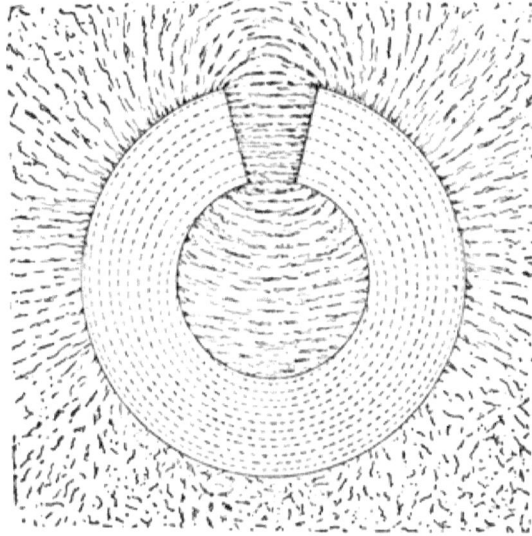

Fig. 73.—Magnetised Iron Ring, as in Fig. 72, but with wider air-gap.

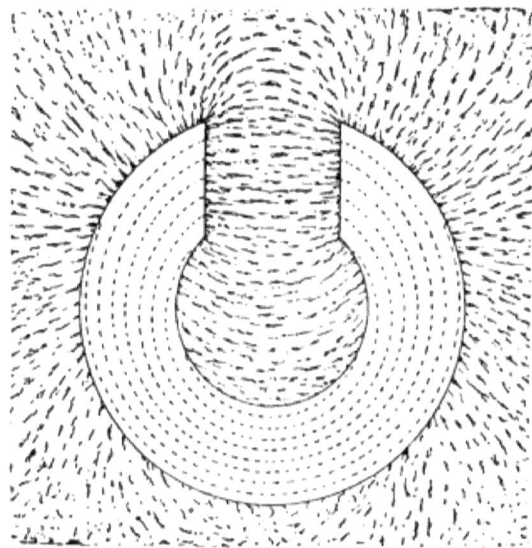

Fig. 74.—Magnetised Iron Ring, as in Fig. 72, but with still wider air-gap.

circuit. This change may be produced in two ways—either by stopping the primary current or by reversing its direction. In the second case the amount of change in the induction is doubled, because the reversing of the primary current amounts to first stopping it, and then starting it again in the opposite direction. Take the first case, viz., the complete iron ring, and imagine the direction of the primary current continually being changed or alternated. This would produce an alternating magnetic flux round the ring, the direction of the magnetisation changing its direction with every change in the direction of the primary current. This changing magnetism passing through the secondary circuit would set up in the secondary circuit an alternating secondary electromotive force, and therefore, if the secondary circuit is completed, an alternating secondary current. It will be at once evident, however, that there are two ways in which we may set up this changing magnetism in that half of the ring on which the secondary circuit is wound. We may either employ the undivided ring, and keep continually changing the direction of the primary current in the primary circuit, or we may maintain the primary magnetising current constant in one half of the ring and turn round that half of the ring on which the secondary circuit is wound, so as to keep on reversing the direction of the magnetic current through the secondary circuit. A little consideration of these two modes of producing the changing induction through the secondary circuit will show that they really amount to the same thing.

In the light of the above remarks it will be found very easy to understand the general action of the alternate current transformer and dynamo. It has been explained in the previous Lectures that the power conveyed by a continuous electric current is measured by the product of the current strength measured in amperes and the difference of pressure or potential between the ends of the circuit measured in volts. The same amount of power, say one horse-power, can, therefore, be conveyed, either in the form of a large current having a small fall in pressure or in the form of a small current having a large fall in pressure. We are familiar enough with this effect in the case of water. A waterfall may consist of a very large body of water falling down a very moderate height, such as the well-known falls of the Rhine at Neuhausen, in Switzerland; or it may present itself in the form of a much less quantity of water falling from a much greater height, like the falls of the Anio, at Tivoli, in Italy. In either case the power which the water would be capable of exerting, if utilised by means of turbines or waterwheels, would depend, not on the height of the fall nor on the quantity of the water flowing down, taken alone, but on the product of the numbers representing respectively the height and the quantity. The height of the fall determines the velocity or pressure of the water at the base of the fall. It is impossible to get any work out of water unless we are able to let it fall from one level to a lower one. Hence it is that stores of water at a height above the level of the sea represent available sources of energy.

The power which originally lifted the water up before it can fall must have been the evaporative power of the sun's heat. We commonly speak of using *water-power*; but, as a matter of fact, we are really employing in these cases *sun-power*. The falls of Niagara represent the accumulated water drainage of half a continent falling over a precipice of 150 feet in height on its way to the sea. This water, however, was originally evaporated from lakes, rivers, or the sea, before it could be precipitated as rain over Canada and North America. The power which lifted these water molecules from the surface of the Atlantic or Pacific Oceans into a position in which we are able to make them give back part of their energy of position was the radiation proceeding from the sun. The dynamos now being placed in position to tap off some of the available power of the Falls of Niagara will in reality be driven by sun-heat, radiated to the earth at some previous time.

These hydraulic facts have an exact parallel in electrical science. A current of electricity can only do work by falling down from one electric pressure or potential to a lower one, and the rate at which it can do work is measured, not by the current strength or by the fall in potential alone, but by the product of the numbers representing these two quantities. A current of electricity of 10 amperes falling in pressure by 1,000 volts, represents the same power of doing work as a current of 1,000 amperes falling 10 volts in pressure. There is, however, a considerable difference between the two agents. The small current of high pressure is very different in many respects from the large current of low pressure. In the first place, the high-pressure current is more dangerous to handle or deal with, and requires much better insulation than the low-pressure current. On the other hand, when a current of electricity flows through a conductor a part at least of its energy is frittered away into heat by reason of the necessary resistance of the conductor. The amount so dissipated is, by Joule's law, measured by the product of the square of the current strength and the resistance of the conductor measured in ohms. Hence, if flowing in conductors of the same size in cross-section and length, a current of 1,000 amperes would generate 10,000 times more heat per second than a current of 10 amperes. Suppose, then, that we wish to convey a power of 10,000 watts to a distance, we can do it either by using a current of 10 amperes flowing out and back along electric mains between which 1,000 volts pressure is maintained, or by using a current of 100 amperes flowing in mains between which 100 volts pressure difference is preserved. If, however, we wish to dissipate or waste only the same fraction of our 10,000 watts power, say 10 per cent., in conveying it, we shall have, in the case of low pressure current, to employ electric mains of 100 times less resistance in the second case than in the first. In other words, we must provide a conducting channel for the large current of 100 amperes which shall be 100 times the cross-section, and therefore 100 times the conductivity, of that which it would be necessary to lay down for conveying the small current of 10 amperes, with the same proportionate loss of energy. It will, therefore, be evident that, as far as mere cost of copper is concerned, it is cheaper to convey electric energy in the form of small electric currents at high pressure than as a large electric current at lower pressure. On the other hand, the high pressure conductors cost more to insulate, so the advantage is not simply proportionate to the economy in copper.

The point to which I wish now to direct your attention is that we have in the *transformer* an appliance which enables us to convert small currents of high pressure into large currents of low pressure. In so doing we do not create any energy—in fact, we waste some of it; but the transformer is a device which enables us to change the form of electrical power just as a pulley or other simple machine enables us to change the form of mechanical power. By the use of a pulley and tackle a man exerting a small force through a great distance can raise a very heavy weight through a smaller distance.

The alternate-current transformer is an appliance which operates on electric power just as a lever or pulley does on mechanical power. It consists of an iron ring or circuit of some form or other which is wound over with two wires, a thick wire and a thin one. These circuits are carefully insulated one from the other. The thin wire generally makes some ten or twenty times more turns round the iron ring than the thick one. If a small alternating current of electricity from a high pressure source is passed through the thin wire, it creates an alternating magnetism in the ring of iron, magnetising it first one way and then the other in a direction round the axis of the ring. This changing magnetism creates, as before explained, another alternating electric current in the thick-wire circuit, called a secondary current, and this current can be made to have a lower pressure and greater strength than the primary current by suitably proportioning the number of windings of the two circuits. Hence we can transform a small current produced by a pressure of 2,000

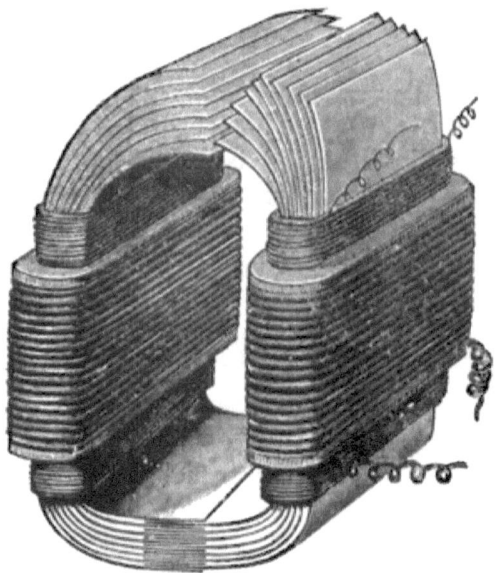

FIG. 75.—One form of an Alternate-Current Transformer in process of manufacture.

volts into a current nearly 20 times as strong, but having a pressure only of 100 volts. In this transformation there is a certain loss of power, or dissipation of energy, but in good transformers this will not exceed from 1 to 4 per cent. of the amount of power, transformed. We can, then, by means of

ELECTRIC DISTRIBUTION. 213

the transformer, carry electric power at high pressure and transform it down to a low pressure at the place where it is to be used. In transformer systems of electric supply this

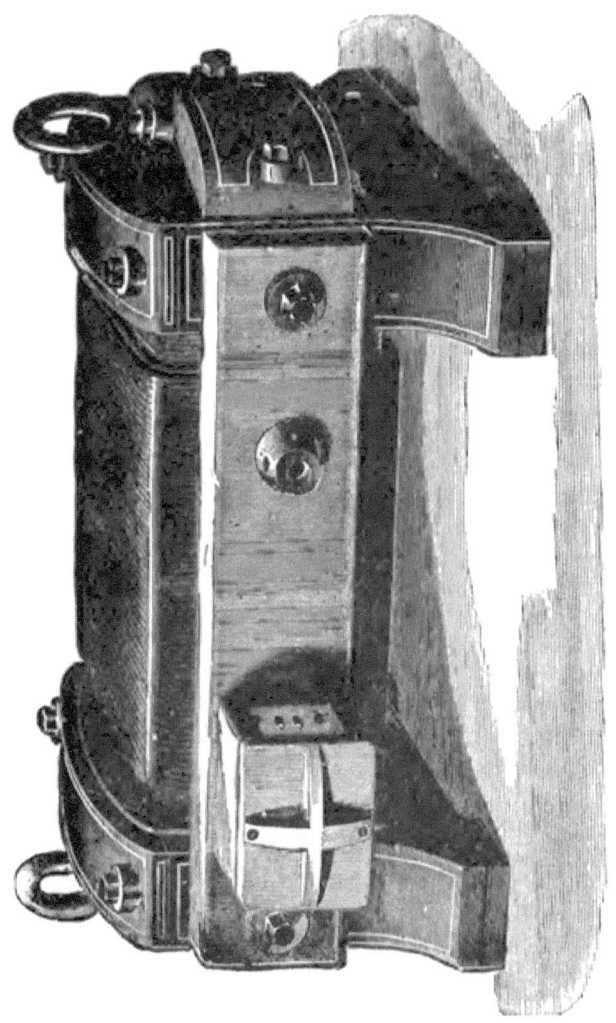

FIG. 76.—One form of an Alternate-Current Transformer, complete.

is the method adopted, as opposed to the other or direct-current supply, in which continuous currents flowing uniformly

in one direction are used, and which distribute current at
a lower pressure for use without transformation.

The practical construction of the transformer is carried out
as follows :—A number of thin iron bands, strips, or plates are
so arranged as to form an iron ring, or its equivalent. This
ring is then wound over with two circuits of insulated wire,
one a fine wire, called the primary circuit, and one a thick
wire, called the secondary. In Fig. 75 is shown such a transformer in process of manufacture.

The transformer, when completed, is generally enclosed in
an iron case, and is provided with proper terminals for the
two circuits (*see* Fig. 76). Transformers made as above are
generally constructed for various transforming powers, such
as 1 horse-power (H.P.), 10 H.P., &c. This means that the
transformer, say of 10 H.P., can transform the energy of a
current of about 4 amperes at a pressure of 2,000 volts into
energy in the form of a current of about 74 amperes at a
pressure of 100 volts. The sizes of transformers are often
given in *kilowatts*, and then we have to remember that one
kilowatt is equal to about $1\frac{1}{3}$ H.P. Hence a 30-kilowatt transformer is a 40-H.P. transformer. It will, therefore, be seen
that a transformer is only a development of the simple iron
ring wound over with two circuits, as described at page 180,
and illustrated in Fig. 69. The object of building up the
iron core, or ring of iron strips or plates, is to check, as far
as possible, the production of electric currents in the mass of
the iron core, which, if unprevented, would be a source of
waste of power.

We are also able to present a similar simple explanation of
the mode of action of the dynamo machine. Referring again
to Fig. 70, we see that we can reverse or change the direction of the magnetic induction or current flowing through

the secondary circuit by making two cuts or air-gaps in the iron ring, and then rotating that half of the ring on which the secondary circuit is bound. This can, perhaps, be better understood by reference to another diagram (*see* Fig. 77). In this figure the shaded horse-shoe shaped part marked M is supposed to represent an iron core, which is magnetised by an electric current passed through a wire wound round that core, but which wire, for the sake of simplicity, is not shown on the diagram. This magnetic circuit is completed by placing another iron mass, represented in the figure by the shaded rectangle in between the poles or ends of the horse-shoe magnet. We thus obtain an iron circuit having two cuts or air-gaps in it, corresponding with the split ring shown in Fig. 70.

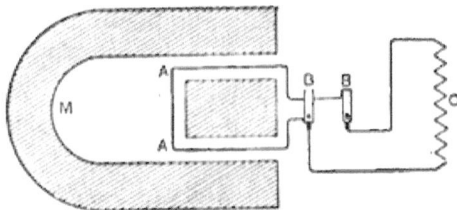

Fig. 77.—Skeleton Diagram, showing the principle of the Dynamo Machine.

Suppose, now, a magnetic induction or magnetic flux to exist in one direction round this air-iron circuit, and to be produced by a primary current flowing in a primary circuit of insulated copper wire which is wound on the horse-shoe shaped iron core. This magnetic induction or flux will pass across the air-gaps, producing there a powerful magnetic field. Next, let us suppose that the iron block, or, as it is called, the armature core, is wound over with a circuit of wire indicated by the rectangle A A in Fig. 77. Let this core and armature wire be capable of rotating round an axis, and suppose that, by a couple of contact rings B B, we can always keep the ends of the armature circuit A A in conducting connection with an external circuit **C**. If the rectangular core and its wire winding is made to revolve round its axis, it **is easy to see that the direction of the magnetic flux** or induction through this secondary or armature circuit will be continually reversed, and hence that an induced secondary current will be created in it, which will also be continually reversed—that is, it will be an alternating current.

A dynamo, therefore, consists essentially of an iron core, or *field magnet*, as it is termed, which is wound over with a primary circuit or field wire circuit, and this circuit, when traversed by an electric current, creates a powerful magnetic field in the space between the poles or ends of the field magnet. In the interpolar space is placed an iron cylinder or ring, called the *armature core*, and this serves to complete the magnetic circuit. On this armature core is wound another coil or coils of wire called the armature windings, which correspond to the secondary circuit of the transformer. The armature core and its windings are rotated by being fixed on a spindle to which is usually attached a pulley. In Fig. 78 is shown a representation of an Edison-Hopkinson dynamo, in which the field magnets and armature are easily distinguished. In one large class of dynamos, called alternate-current machines, or alternators, the secondary current which is sent out from the secondary or armature circuit is, as observed above, an alternating current.

For many purposes, however, it is essential to produce a continuous current, or a current always in one direction. To obtain this, an additional organ has to be furnished to the dynamo. This is called a *commutator*, the function of which is to convert the alternating current produced in the armature coils into a continuous current in the external circuit. A little attentive consideration will, however, show that a dynamo and transformer are essentially the same appliances in nature, only that in the latter the necessary reversal of the flow of magnetism through the secondary circuit is obtained by alternating or reversing the

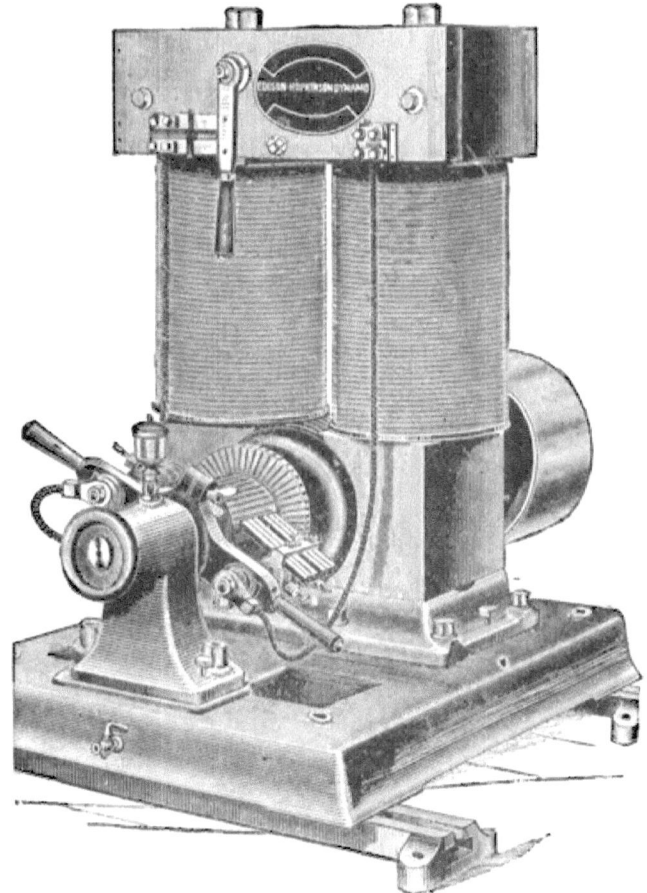

FIG. 78.—Edison-Hopkinson Dynamo.

direction of the primary current, whereas in the case of the dynamo the necessary reversal is obtained by rotating or turning round that part of the iron magnetic circuit on which the secondary circuit is wound. This rotation of the armature of the dynamo may, in practice, be effected either by putting a pulley on the shaft and driving it by a belt or rope, or by coupling the dynamo shaft directly to the shaft of a high-speed steam engine.

One of the methods of direct driving, as this is called, most widely adopted in England is to employ a high-speed Willans engine, the shaft of which is continuous with that of the

FIG. 79.—View of an Alternator direct driven by being coupled to a Willans Engine.

dynamo. In Fig. 79 is shown an illustration of such a combination. It is obvious that this arrangement is exceedingly economical of space, and hence has been widely adopted in the arrangement of electric generating stations, in which space is of importance.

It is impossible, as observed at the beginning of this Lecture, to do more than give here the briefest outline of the different methods employed for distributing current for illuminating purposes; but the non-technical reader will probably be able to grasp most easily the nature of the various systems employed by my selecting one or two typical instances, and describing these somewhat in detail. At the present moment there are two principal systems of electric lighting, both of which have their advantages: these are called respectively the alternating current supply system and the continuous current supply system. In the first of these a generating station is established, which is provided with dynamo-electric machines called alternators, which are capable of producing an electric current alternating in direction a large number of times in a second. In the case of alternating currents the current flows in one direction in the circuit for a short fraction of a second, is then reversed, and then flows in the opposite direction for a short fraction of a second, and repeats the same cycle of operations continuously. The number of times per second which the current cycle is repeated—namely, a flow in one direction succeeded by a flow in the opposite direction, like the ebb and flow of the tide—is called the *frequency* of the electric current, and the frequencies most usually employed are either 40, 83, 100 or about 125. In England the practice generally is to use a frequency of 80 or 100, on the continent of Europe the lower frequency of 40, and in America the higher of 120. This alternating current is generated by machines called alternators at a high pressure, which is generally either 1,000, 2,000 or 2,400 volts, and in some cases 5,000 or 10,000. This high-pressure alternating current is then led out from the station by highly insulated conductors, which are called the primary mains, and these primary mains are generally made in what is called the concentric form. A stranded copper cable is covered over with insulating material, and then a circular strand of wire is plaited over the insulating material, and this again is protected by a further layer of insulation, and finally by a covering of steel wires, called the armour. In Fig. 80 is shown a section of this steel-armoured concentric cable.

This armoured cable is laid down in the streets under the pavements, and is then usually conducted to certain places which are called *transformer sub-centres*. These are small rooms, either excavated out underground or built above ground, and to them the high-pressure primary

Fig. 80.— Cross-section of a Concentric Lead-covered Steel-Armoured Cable, for High-Pressure Transmission. The central white dots represent the inner conductor, and the outer circle of white dots the several wires composing the outer conductor. The shaded portion shows the outer covering of lead.

current is brought by these "primary" cables. In these transformer sub-centres are placed a number of transformers, which, as already explained, are connected with the primary mains, so that all their primary circuits are in parallel across the mains. The secondary circuits of the transformers are then connected to another set of underground mains, which are called the secondary distributing mains, and these distributing mains are laid down in the streets which are to be furnished with light. It is most usual to employ a transformation ratio of 20 to 1, so that the pressure of 2,000 volts is reduced to 100 volts at the other side of the transformer.

In many cases the secondary distributing system is laid down on the three-wire system (*see* Fig. 81). In the transformer sub-centre are placed a number of transformers, T T, which have primary and secondary switches, by which their primary circuits can be connected to the primary mains M M, and their secondary circuits to the secondary mains A B C. During the

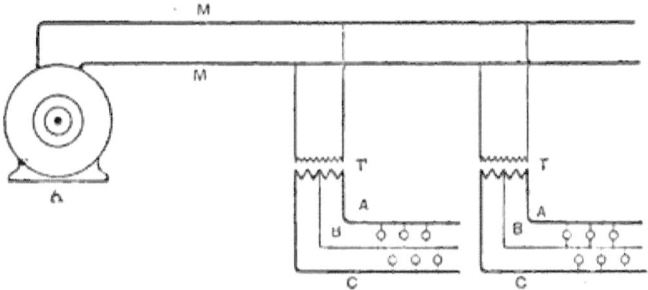

Fig. 81.—System of Distribution by Alternating Currents and Three-wire Secondary Circuits. A is the Alternating Current Dynamo, M M the Primary Mains, T T the Transformers, and A B and C the Three-wire Secondary Circuits with Lamps across them.

hours of small demand (or "light load," as it is technically called) only one of these transformers, which is called the master transformer, is kept connected to the system; but during the hours of heavy demand (or "full load") two or more transformers are added on, so as to share the load. In some cases of very scattered lighting, instead of collecting together the transformers in sub-centres, one or more transformers are placed in each house or building which is to be supplied with current.

In such a transformer system the sources of losses of energy are as follow: There is a certain constant loss of energy in every transformer which is kept connected to the primary mains, owing to the fact that the incessant reversal of the magnetisation of the iron core of the transformer necessitates an expenditure of energy. This is called the "open circuit loss" of the transformer, and in thoroughly efficient transformers does not exceed from one to two per cent. of the whole of the nominal output of the transformer. Thus, for instance, if the transformer is one which is capable of transforming 10 horse-power of electric energy from the form in which it exists as a current of about four amperes flowing under a pressure of 2,000 volts into electric energy which exists in the form of 75 amperes at a pressure of 100 volts, such a transformer can be made to dissipate or waste only about one-seventh of a horse-power in the continual magnetisation of its iron core. When the transformer is doing its full work, or is loaded up to any extent, in addition to this open circuit loss—which is also called the iron loss in the transformer, because it takes place in the iron core—there is an additional loss, which is called the "copper loss," and this is due to the heating effect produced by the currents in the primary and secondary copper coils of the transformer. In well-designed transformers the copper loss at full load is about equal to the iron loss at no load, and it will therefore be easily seen that the result is to give the transformer an "efficiency," as it is called, of about 97 per cent. at full load—that is to say, of the whole power supplied to the transformer in the form of high-pressure current 97 per cent. is given out by the transformer in the form of low-pressure current. It is possible to construct a transformer so as to have as much as 90 per cent. efficiency at $\frac{1}{10}$th of full load. The most important point to consider, however, is, not the efficiency of the transformer at full load, or its efficiency at light loads, but to consider the so-called *all-day efficiency* of the transformer—that is to say, if the transformer is employed for 24 hours, doing such lighting as may be demanded of it, the important quality which it should possess is that of having a high all-day efficiency; in other words, the ratio between the total number of units of electrical energy which are taken out of the transformer must bear a large proportion to the number put into it.

As already explained in the Lecture on "Electric Glow Lamps," the ordinary demands for lighting in residential or public buildings are of so varied a nature that the full current for the whole of the lamps placed in them is only required for a very small portion of the 24 hours. For instance, in an ordinary private residence, during the early part of the day there may be a small demand for lighting in the lower part of the house; during the greater part of the day very little; in the afternoon some more lamps will be turned on; in the evening (in the winter between 4 and 10 or 11) there will be a variable but much larger demand; and during the night very little. If on a piece of paper a line is taken and divided into twenty-four parts, and a perpendicular drawn at each of those points representing by its altitude the number of amperes of current taken by the house at that moment, a curve joining the tops of all these lines gives us the *load diagram* of the house, and the area of this load diagram gives the whole quantity of current taken by the house. If, instead of erecting perpendiculars proportional to the current, we erect perpendiculars which are proportional to the power measured in watts being taken up in the house at that time, we get the *watt-hour* diagram, and the whole area of this diagram measures the total amount of energy in Board of Trade units taken up in the house during the twenty-four hours. The ratio between this amount and the full amount which would be taken if all the lamps were kept burning for the whole period is called the load factor of the house, and the load factor of the house is on an average not more than 10 per cent.

Accordingly, if a house or a series of houses are supplied from one transformer centre, the actual demand at any one moment for electric power may be very small; and the important point with regard to transformers is that they should possess a very high efficiency at low loads, in order that, on the whole, the number of units of electrical energy that are wasted in the transformers in magnetising the iron cores may not be a very large proportion of the total energy which is supplied to the houses. Transformers can now be easily made to have an 80 per cent. all-day efficiency, and, under these circumstances, the total amount of energy supplied to the customers may be, and often is, as much as 80 per cent. of that which is sent out from the station. In inferior systems of alternating current supply the efficiency of distribution may fall very far short of 80 per cent., not amounting to even more than 50 per cent.

Having placed these general principles before our minds, we can now proceed to consider the main features of an existing alternating current supply system, which has been worked out with great care and forethought, namely, that adopted for the supply of electric energy for the lighting of the city of Rome. The electric lighting of Rome is conducted from two electric lighting stations, one of which has been established in the gas works at Rome, at the foot of the Palatine Hill, and the other at Tivoli, eighteen miles from Rome. The first station was put down in the year 1889 with the object of supplying current for incandescent and arc lighting within the city. This station occupies the site of the old Circus Maximus, and at the present moment it supplies an equivalent of nearly 40,000 10-c.p. incandescent lamps. The station itself is a substantial brick building in close contiguity to the gasworks. It contains six Ganz alternating current dynamos—four of 600 horse-power and two of 250. Each alternator consists of a series of armature

ELECTRIC DISTRIBUTION.

coils which are contained on the inside of an iron ring frame. A central wheel, which constitutes the flywheel of the engine

FIG. 82.—600 horse-power Alternator in the Cerchi Electric Station at Rome.

FIG. 83.—600 horse-power Alternator in the Cerchi Electric Station at Rome.

driving the dynamo, carries on its external surface a series of insulated coils, which are called the field-magnet coils. The engines are coupled direct to this flywheel, the cylinders of the engines being one on either side. In Figs. 82 and 83 are shown views of these 600 H.P. alternators with the direct-coupled engines. Smaller alternators of 250 H.P. are used during the daytime, or at a time when the demand for lighting is not great. These alternators generate an alternating current at a frequency of 40 and a pressure of 2,000 volts.

Fig. 84.—View of the Hydraulic Tower at Tivoli.

They send their current into three primary cables or feeders, which are concentric steel-armoured conductors, and these feeders deliver current to certain transformer centres and to transformers which are placed in the various buildings to be lighted. The total length of mains laid down is about 25 kilometres, or 16 miles. At the end of 1892 there was a total number of lamps connected to the system equal to 34,464 10-c.p. lamps. In the transformer centres and in the houses are placed transformers for reducing the pressure from 2,000 volts to 100 volts, two sizes of transformers being mostly used, one capable of transforming power equal to 10,000

Fig. 85.—View of Power Station at Tivoli.

watts and the other equal to transforming power of 5,000 watts. From these transformers secondary cables are laid, distributing secondary current at a pressure of 100 volts to various buildings.

The capacity of this station being fully exhausted, it was determined to furnish an additional supply by means of a second station, and to take advantage for this purpose of the large available water supply of Tivoli. The village of Tivoli, which occupies the site of the ancient Tibur, stands on a spur of the Sabine Hills. The beauties of Tivoli and its surrounding country were celebrated in undying verse by Horace, whose Sabine farm was not very far distant. The chief natural attraction is the fine cascade formed by the River Anio, one fall of which is 340 feet in height. Here also in classic times stood Hadrian's villa, of which the grounds once covered an area of several square miles, and contained an unrivalled collection of works of art. Upon this romantic spot the modern engineer has laid his hands, and has compelled a large portion of the power running to waste in these waterfalls to be directed to the purpose of electrically lighting modern Rome. With this object, a hydraulic canal was first constructed, leading from one of the upper reaches of the River Anio to the top of a tall tower (see Fig. 84). This tower contains an iron hydraulic fall tube six feet in diameter. At a distance of 150 feet below the top of this tower, and about half-way down the side of the hill a power house (see Fig. 85) has been constructed, into which the termination of this hydraulic fall tube is brought. The water from the upper level conveyed by the hydraulic canal falls down this hydraulic main, and the tube has a capacity for delivering 100 cubic feet of water per second at a pressure equal to that due to a height of 150 feet, which is about equal to 60 lb. on the square inch. The end of this hydraulic tube terminates in three lateral branches, each three feet in diameter, and closed with a hydraulic valve capable of resisting the enormous pressure of the water above. Each branch of the main also sends off three subsidiary branches, which are led to three Girard turbines or hydraulic motors. These machines are practically water engines, in which the flow of water is made to cause the revolution of a wheel with curved blades which is enclosed in an iron case. In one room of the power house there are nine of these turbines, six of 350 H.P. and three of 50 H.P. The water that passes through these turbines is delivered into a tailrace, and then returns back again into the Tivoli Falls at a lower level. Coupled to each of those turbines is a Ganz alternating-current dynamo. Six of these dynamos are of 350 H.P. each, and can furnish an alternating current of 45

FIG. 86.—Turbine and Exciter Dynamo in the Tivoli Station. Tivoli-Rome System.

amperes at a pressure of 5,000 volts; three of the dynamos are direct-current machines of 50 H.P., and can furnish a continuous current of 150 amperes at a pressure of 125 volts. These smaller dynamos are employed for furnishing the current required to magnetise the field magnets of the larger machines. A view of one of these smaller machines and its associated turbine is shown in Fig. 86. These six large alternators send their current to a switchboard which is 60 feet long and 15 feet high. From the power house proceed four stranded copper cables, each consisting of 19 copper wires, the over-all diameter of each cable being about one inch. Each cable weighs 980 kilogrammes per kilometre, or nearly two tons per mile. These cables convey the current across the Campagna from Tivoli to Rome, a distance of about 18 miles, and in all four lines there is a total weight of about 120 tons of copper. These four cables

Fig. 87.—The Half-way House. Tivoli-Rome Electric Lighting System.

are supported by porcelain insulators, which are carried upon crossbars on iron posts, placed at intervals throughout the whole length of the line. These porcelain insulators are similar to those used for carrying telegraphic lines, with the addition that they contain an arrangement for keeping the edges of the insulator moistened with a highly insulating oil. These oil insulators have been designed specially with the object of insulating the line for the 5,000 volts pressure.

Half-way between Tivoli and Rome there is a half-way house (*see* Fig. 87), in which dwells a custodian whose duty it is to inspect the whole of the line. The power station at Tivoli provides in all a total of 2,000 H.P. in the form of an electric current at a pressure of 5,000 volts. The copper transmission cables are made of hard-drawn copper wire, and of a cross-section of 100 square millimetres. The poles carrying the cables are placed 35 metres apart,

FIG. 88.—View of the Switchboard Arrangements in the Porta Pia Transformer House. Tivoli-Rome System.

in almost a straight line across the Campagna, and there are 707 poles altogether between Tivoli and Rome. The resistance of each cable is about 38 ohms for the eighteen miles. Each cable can convey 100 amperes, and the four cables—that is, the two pairs—can therefore convey 200 amperes from Tivoli to Rome, which is equal to the output of five of the large alternators. In the passage from Tivoli to Rome the current undergoes a fall in pressure due to the resistance of the line, which is about 800 ohms in the eighteen miles when working at full load. The cables, therefore, dissipate about 16 per cent. of the whole of the power capable of being transmitted at full load. The alternators have been so constructed that they can all work in parallel, sending their current into one or both pairs of cables as may be desired. On arriving at Rome the current enters a transformer house (see Fig. 88) placed just outside the Porta Pia of Rome. In this transformer house are placed a series of transformers, sixteen in number, each capable of transforming 30,000 watts from 5,000 to 2,000 volts. This transformer house is a substantially-built stone building two stories in height. The current, reduced from 5,000 to 2,000 volts, is then transmitted by underground cables through the streets of Rome to certain transformer centres, where it is again reduced in pressure to 100 volts. In addition to this, a portion of the current, at a pressure of 2,000 volts, is utilised for working a series of alternating current arc lamps. There are six circuits of these arc lamps, each circuit being arranged for 48 lamps. Each series of lamps takes a current of 14 amperes, and there is a special automatic pressure-regulator attached to the transformer supplying these arc light circuits, so that the current is kept perfectly constant, whatever may be the number of lamps upon the circuit. In addition to the arc light circuits, there are five or six incandescent lighting circuits consisting of Siemens concentric steel-armoured cables laid underground. The current supplied from Tivoli can be arranged to assist the Cerchi station, so that the two stations, 18 miles apart, one in the gasworks and the other at Tivoli, assist one another in furnishing current for incandescent and arc lighting in Rome as may be required. This Tivoli station is one of the finest examples of an alternating-current station operated by water-power, and it has been in perfectly successful operation since the year 1890. In order to protect the overhead line, going over 18 miles along the Campagna, from damage from lightning there are lightning protectors of the kind described on page 157, placed at Tivoli, in the transformer house at Rome, and in the half-way house shown in Fig. 87. These protectors have proved sufficient to guard the dynamos and transformers from danger by lightning stroke.

The general arrangement of a low-pressure continuous current station on the three-wire system may be described as follows: In the station are placed a series of dynamos, each dynamo in most English practice being direct-coupled to a high-pressure compound steam engine. In Fig. 79 is shown

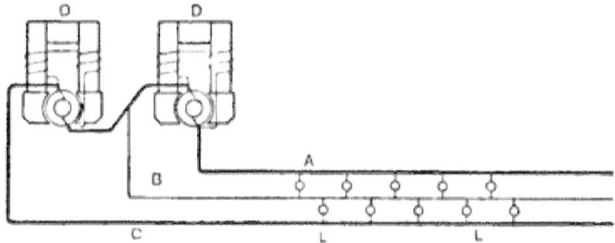

FIG. 89.—Diagram showing the arrangement of Dynamos D D, and Lamps L on the "Three-wire" System of Distribution.

such a steam dynamo, in which the armature shaft is coupled direct to the crank shaft of a Willans engine. This combination of engine and dynamo on the same bedplate is obviously very economical in floor space, and has therefore naturally become very popular in places where space is valuable. A series of these engine-dynamo sets is generally arranged in a central station, each set being either similar in size and appearance, or else in graduated sizes. In Fig. 91 is shown the interior of the St. Pancras electric lighting station, and in Fig. 92 the interior of the central station at Glasgow. In each of these the steam dynamos are arranged to generate current generally at either 220 volts or 440 volts, delivering this current to a switchboard. If the current is supplied direct to the circuits, then the dynamos are joined in pairs between the three bars of the switchboard, as shown in Figs. 89 and 90. From the ends of these "omnibus bars," as they are called, are laid out a series of feeding mains or feeders, which feeders terminate in the distributing mains laid down in the streets of the district to be supplied. The circuits of the various houses are connected up to these triple distributing mains in such a manner that as nearly as possible one-half of the lamps are joined in between one pair of mains and the

Fig. 90.—Dynamos in a Three-wire Low-Pressure Station.

other half between the other. In laying out such a district great discretion and knowledge have to be brought to bear in selecting points, which are called the feeder centres, so that at those points the pressure may be kept as constant as possible. Each dynamo in the central station may be regarded as a pump, pumping its current into the positive main omnibus bar, from which it flows out by the feeders to the distributing circuit and returns back again to the opposite main. During the hours of heavy demand for current there is, of course, a large current flowing out by these feeders, and a correspondingly large fall of pressure down the feeder. Hence the pressure at the terminals of the dynamos has to be kept up in order to allow for this fall. It is usual to maintain the pressure at certain feeding points constant at 100 or 110 volts, whatever may be the nature of the supply, and to do this the pressure at the dynamo terminals has to be varied from 110 to 150 volts or thereabouts, as the demand varies at different hours of the day or night. It will thus be seen that in the feeders in a low-pressure station the loss of energy is greatest at the time of greatest load, because they are then traversed by the largest currents, and, therefore, the total supply of energy from the dynamos has to be equal to the amount required by the lamps in use plus that required to supply the corresponding loss in the feeders. A low-pressure continuous current station has, consequently, its least efficiency of distribution at the time of fullest load. Exactly the opposite is the case with the alternating current supply. In this case the efficiency of distribution is a maximum at the time of largest demand.

We may take as a good example of a low-pressure station the St. Pancras Vestry Electric Lighting Station, which has been carried out by the local authority of the parish of St. Pancras, in the north-west of London, for the supply of electrical energy for lighting and motive power purposes in that district. In this part of London, which is densely populated, it was considered advisable to adopt the continuous current low-pressure system, and a typical station of this kind was, therefore, designed for the district by Prof. Henry Robinson. The following is a description of the works as given by him: The station buildings consist of an engine-house 106ft. by 26ft., a boiler-house 53ft. by 35ft., a coal store 43ft. by 11ft., a battery room 40ft. by 14ft. 6in., as well as a testing-room, office, stores, and an underground tank for condensing water capable of containing 170,000 gallons of water, and a chimney shaft 5ft. square inside and 90ft. high. The dynamo room contains eleven engines and dynamos, which are erected on a concrete foundation, surrounded by sand to prevent vibration being communicated to the walls. The floor is carried independently of the engine foundations by cantilevers from the walls. The engines employed are of the Willans compound

Fig. 91.—View of the Interior of St. Pancras Electric Lighting Station, London, showing the Dynamos.

central valve type, and the dynamos are of the 6-pole Kapp type made by Messrs. Johnson and Phillips. A view of the interior of this station, showing the long perspective of engines and dynamos, is given in Fig. 91. Each Willans engine is coupled direct to its corresponding dynamo. These dynamos furnish a continuous current, nine of them giving 680 amperes at a pressure varying from 112 to 130 volts, and three of them giving 145 volts, with a small current for charging secondary batteries at a distant sub-station. The remaining two dynamos are wound for an output of 90 amperes at 510 to 575 volts. These are used for working the street arc lamps as well as for charging in series four sets of storage batteries at the central station which are capable of discharging at a rate of 60 to 75 amperes. The boiler plant consists of five Babcock-Wilcox boilers, each capable of evaporating over 5,000lb. of water per hour, the working pressure being 170lb. on the square inch. Along the top of these runs a steam main into which all the boilers deliver their steam and from which the engines separately take their steam. The steam, having done its work in the engines, is then either delivered into the atmosphere or passes into a jet condenser, which draws its water from the bottom of the underground tank. From the top of this tank the hot water is pumped by independent pumps to a cooling apparatus, which is capable of dealing with a minimum of 10,000 gallons of water per hour. The cooling arrangement consists of a large surface of corrugated sheet iron attached to a framing placed round the chimney. The water, after flowing along the corrugated surface, is collected underneath and returned to the bottom of the tank. There is also an air pump in the station for delivering dry air into the street main culverts at the rate of 5,000 cubic feet of air per hour.

The street lamps are Brockie-Pell arc lamps, placed on tall iron posts placed in the centre of the road where admissible. They are fixed at distances varying from 160ft. to 245ft. apart, the height of the lamp varying from 22ft. to 25ft. above the pavement. These lamps are worked 11 in series near the central station and 10 in series at a distance, and each lamp takes a current of 10 amperes, which is supplied by the two 500-volt dynamo machines. The main switchboard for the incandescent lighting is arranged on the "three-wire" system. Each dynamo is provided with a double-pole switch and copper coupling strips to connect the machines to the third or middle wire and to the different bars on the feeder board on either side of the circuit. There is an amperemeter on each dynamo to indicate the current being given out. On the switchboard there are four positive omnibus bars and four negative bars, and at full load one machine can be run on to each bar and be worked at any pressure which may be necessary to serve the feeders which are switched on to it. Thus the dynamos may be worked at different pressures to suit the demand in the districts which they serve.

There are seven feeder mains going out of the station, in addition to a direct supply to the distributing mains. The mains are laid throughout the district on the three-wire system, the principal mains being sufficient to supply 25,000 incandescent lamps of 16-candle power in use simultaneously. These conductors are laid down under the streets, and are composed of copper strips $1\frac{1}{2}$in. wide and $\frac{1}{4}$in. thick, supported on edge in glazed porcelain insulators, which are carried on small cast-iron brackets built into the walls of the culverts. Some of the mains are cables laid in cast-iron pipes, and some of them are armoured cables. In addition to this, the current is supplied from the central station to a distant secondary battery station 1,140 yards away, which contains two batteries of 58 E.P.S. cells each. By varying the number of dynamos in use at the station and the number of feeders in connection with the omnibus bars, the engineers in charge of the station have it in their power to regulate the electric pressure between the distributing mains laid down in the streets, and the supply is so regulated as to keep the pressure between the middle main or wire and each of the outer wires to 110 volts.

A very similar station was established for the supply of electric current in Glasgow by the Glasgow Corporation. A view of part of the interior of this station is shown in Fig. 92.

ELECTRIC DISTRIBUTION.

The station buildings are substantially built, standing on a layer of concrete two feet thick laid throughout the entire block. These buildings consist of an engine and dynamo room, boiler room, workshop, stores, and offices. The boiler room contains six steel boilers of the marine type, measuring 12ft. by 10ft. in diameter, and working at a pressure of 160lb. Each contains two furnaces of the corrugated type, with an internal diameter of 3ft. These boilers deliver steam through a ring

FIG. 92.—View of Part of the Glasgow Corporation Electric Lighting Station.

main into the engine room. The engine room contains a series of Willans engines coupled to dynamos (see Fig. 92). The engines are Willans central valve compound engines, and the nine dynamos are two-pole shunt-wound machines with drum armatures. The boilers are fed with gas coke obtained from the Corporation gas works. Over the boiler house is placed an accumulator room containing 114 secondary battery cells. Each cell contains 61 lead plates, and has a capacity for storing up and discharging 1,000 ampere-hours of electrical quantity. The current from the dynamos, which is generated at from 210 to 230 volts, is led to a distributing board in the station, and this supply has a three-wire system of distribution laid down in the streets. The three copper conductors consist of copper strip, which is laid on porcelain insulators contained in iron culverts. The current is brought to the two outer wires of the three-wire strip by conductors, which, as in the St. Pancras station before described, are called "feeders," and which consist of only two mains, a positive and a negative main. The lamps are all joined in parallel between the middle main of the distributing mains and one of the outer ones, the total of about 46,000 eight candle-power lamps being, as far as possible, connected so that one-half of the lamps are joined in between the middle main and one of the others, called the "positive," and the other half of the lamps being connected in between the middle main and the other conductor, called the "negative." The middle wire, therefore, serves to carry the balance of current from one part of the distributing system to the other at those times when the number of lamps joined in between the two sides of the three-wire system is not exactly equal.

These two stations may be taken as fairly typical examples of low-pressure stations as established in Great Britain, although many other equally good, such as the Kensington and Knightsbridge and Westminster stations in London, might be described. Such brief space as we are able to give to this portion of the subject in no way enables us to exhaust the description of the whole of the methods for a direct current low-pressure supply; but the general tendency, at any rate in England, is to adopt the direct-driven type of dynamo, on account of the considerable economy which is thereby effected in space. In some low-pressure stations, where space is not of much importance, it is found convenient

ELECTRIC DISTRIBUTION. 241

to drive the dynamos by means of belts from the flywheels of engines.

FIG. 93.—View of part of the interior of the Brussels Electric Station, showing the Dynamos.

In Fig. 93 is shown a portion of the interior of a Continental electric lighting station—that of the City of Brussels.

This station is constructed to contain plant of 3,000 indicated horse-power, the unit adopted being a 500 horse-power engine driving two 250 horse-power dynamos. Provision is made for placing six such units, one being for reserve. The boiler plant in 1894 consisted of three Babcock-Wilcox water-tube boilers capable of evaporating 7,500lb. of water per hour. The engine room has space for six or more, of the horizontal compound condensing type. The low-pressure cylinders are 40 inches in diameter and the high-pressure cylinders 26 inches in diameter, the length of the stroke being 4 ft. The speed is capable of variation between 62 and 75 revolutions per minute. Each engine is fitted with two grooved flywheels 20 feet in diameter, and weighing 20 tons. These drive the dynamos attached to each engine by means of ropes. The dynamos shown in the illustration are of the four-pole type, shunt-wound, with drum armatures, and with an output capacity of 145 kilowatts. These dynamos supply a three-wire distributing system, the cables being laid down underground in iron pipes.

In addition to the generating plant a storage plant of accumulators is provided, in two batteries of 70 cells each. Each battery is capable of giving a normal discharge of 350 amperes for ten hours. The underground cables consist of copper-stranded cables insulated with india-rubber, which are drawn into iron pipes. This work was designed and carried out by the engineers of the Silvertown Company.

The above descriptions of alternating current and continuous current generating stations will be sufficient to give the student a general view of the methods which are at present (1899) employed for the generation of electric current for illuminating purposes. It would lead us into matters too highly technical to attempt to discuss fully when, and under what conditions, each system finds its best application. Suffice it to say that no one system can be described as the best system; both the low-pressure direct current and the high-pressure alternating current systems have certain peculiar advantages and disadvantages, which have to be considered in the designs for a system of electric distribution. The engineer who is called upon at the present time to design and carry out a system of electric supply for public purposes has to consider a large number of facts which determine the choice that shall be made of the method of supply. The tendency at the present time is to employ alternating currents in the transmission of the power from a distance, and to transform them into continuous currents in the region of supply.

Taking such cases as the central areas in large towns, it will be found that at the present time the demand for lighting will vary from one 8-c.p. lamp to three 8-c.p. lamps per yard of distributing main or of house frontage. In the outer portions of large towns and in smaller country ones the demand for light is much more scattered, and in these cases the only possible system of supply which meets the conditions is the alternating current system. It is possible to some extent to combine together the advantages of the two sytems. The generating plant in the station can be made to furnish a continuous current at a low pressure. This can be used to supply the district within a mile or two miles round the station with current furnished directly from the generating machines. Alternating currents can then be used to transmit power to the distant portions of the area of supply, and the pressure can then be reduced, either in sub-centres or in the houses, by means of transformers, to the normal pressure of 100 volts.

In every case in which an engineer is called upon to advise upon the selection of a system of electric lighting for a town, he has to examine carefully the conditions which prevail, in order to guide himself in making a decision whether to employ an alternating current or a continuous-current system. If continuous-current is employed it may be generated either at high or at low pressure. In the case of the low-pressure continuous current stations, the generating machines are now generally designed to give a voltage from 400 to 460 volts. These machines supply the outer wires of a three-wire system of distribution, and the customers' houses are wired for 200 or 210 volt lamps. The lamps are, as far as possible, divided equally between the two sides of the three-wire mains, but, since it is impossible at all times to preserve a perfect balance in the total number of lamps on either side of the middle wire, it follows that, unless compensation is made in some way, the current per lamp will be greatest on that side which has the fewest lamps, and will tend to destroy them if it is above the normal current. Hence some means has to be found to equilibrate the current per lamp, or, which comes to the same thing, render equal the voltage on each side of the middle main. This is generally done by means of secondary batteries, with or without balancing dynamos.

The secondary battery, as now employed, consists of a series of cells, either glass vessels or lead-lined boxes, in which are a double set of lead plates. These plates are placed parallel to each other and slightly separated. Alternate plates are connected together by a lead bar, and are called respectively the positive and the negative sections. The lead plates now used are generally either *pasted plates* or *formed plates*. The pasted plates consist of lead grids which form a reticulated lead plate, the interstices or holes being filled in with a paste or active material, made by mixing with sulphuric acid, red lead or litharge. The positive sections are generally made with red lead and the negative with litharge. The cells are filled with an electrolyte composed of nine parts of water and one of sulphuric acid, or, in some cases, four parts of water and one of sulphuric acid by volume. When the cell has a current passed through it from the positive to the negative plates, the current breaks up the sulphate of lead which is formed as a result of the action of the sulphuric acid upon the oxide of lead, and finally converts the active material on the positive plate into peroxide of lead, and the active material on the negative plate into reduced or spongy lead. This being the case, there is produced an electromotive force, and the cell contains chemical potential energy stored up in it which can be transformed into the energy of an electric current. Generally speaking, from 8 to 16 watt-hours of energy per pound of plates (positives and negatives together) may be stored up. The electromotive force of each cell when so charged is about 2 volts. To charge the cell requires an impressed electromotive force of 2·5 volts. The useful quantity of electricity which can be stored up in a secondary cell varies from 4 to 8 ampere-hours per pound of plates according to construction. During the process of discharge the chemical difference of the plates diminishes, each plate being more or less converted into a white substance, called sulphate of lead. In the course of the discharge the sulphuric acid is taken up out of the dilute acid, and hence the density of the electrolyte diminishes. In the process of charging, the sulphate of lead on both plates is broken up again and the density of the dilute acid or electrolyte increases.

In certain types of plates, called Planté plates, the active material is not put upon the plates, but is formed out of them by a process of treating the plates. The plates are constructed by building them up of lead strips, or in other ways, so as to expose a very large surface of lead to the acid. The surface of one set of plates is then converted into peroxide of lead, and the surface of the other set of plates more or less into spongy lead by subjecting them to a process called "forming," which consists in sending a current first one way through the cell, and then in the other direction, with intervals of rest. The plates are formed in some electrolyte which has a strong oxidising re-action upon lead, and for this purpose nitric acid may be used in and with dilute sulphuric acid in certain proportions.

There are many varieties of secondary cells, differing in details, but the three principal types in general use may be said to be the pasted or packed grid type, as made by the Electrical Power Storage Company, the modified Planté or formed plate, as made by the D.P. Battery Company, and an intermediate class, called the Chloride storage cell, made by the Chloride Electrical Storage Syndicate. This latter cell is so called because the plates are prepared by first casting tablets of a mixture of zinc and lead chlorides. These are held in position in a lead grid cast round them, formed of antimony and lead. These composite plates, as thus formed, are not, however, capable of being used as storage cell plates. The plates so made are placed in a bath of chloride of zinc and acted upon electro-chemically by being coupled to a zinc plate. The final result of this process is to leave the tablet in the form of spongy lead, or lead in a very finely divided condition. The plates that are to form the positive plates in the storage cell then have their spongy lead converted into peroxide of lead by the action of an electric current passing through a solution of sulphuric acid in which the plates to be formed are placed with companion plates of lead. The result of the operation is to produce two sets of plates consisting of lead grids, one set consisting of spongy lead tablets held in the grid apertures, and the other set consisting of peroxide of lead blocks fixed in the same manner. Sets of the positive (peroxide of lead) and negative (spongy lead) plates are placed in a glass vessel in dilute sulphuric acid, the plates being arranged so that alternate plates are identical and are connected together by a lead band welded to them.

This arrangement constitutes a *secondary cell*, and a series of cells connected together is called a secondary battery. The plates in each cell are prevented from touching each other by separators of ebonite or glass, and are held off the bottom of the cell by ebonite bars.

One great defect of the original packed grid type of plate was a tendency for the active material to fall out of the interstices of the lead plate. This has been greatly remedied by careful manufacture. In some types of packed plate the active material is kept on the plate by enclosing the plate in a jacket made of perforated celluloid. This allows access of the electrolyte, but tends to prevent the falling out of material. The plates can then be placed much closer together, and a compact cell obtained which is useful for traction purposes. In some cases the active material is pressed into grooves pressed or cast in the lead plate.

In all cases in which storage batteries are employed in connection with generating stations, and where weight and space are of secondary importance, it is advisable to employ plates substantially made and separated by a considerable distance. The object being to avoid short circuiting by reason of active material dropping out and connecting adjacent plates, and also to allow free circulation of fresh electrolyte between the plates.

A secondary battery in a generating station is set up in a dry and well ventilated room. The cells are placed on wooden trays filled with sawdust, each tray being supported on four glass insulators having oil in them to prevent surface leakage. The cells are carried on stout wooden stands in two or more tiers, and are connected in series with one another. In the case of a three-wire system of distribution two batteries of cells are employed, one on each side of the middle wire.

If, for instance, the system is a three-wire system supplying customers at 200 volts on each side, then it is usual to put in 110 or 120 cells on each side between the middle wire and each outer main. In order to charge these cells an electromotive force of 2·5 volts per cell is required. Hence, if the cells are to be charged at any time, it is requisite to provide a machine called a "booster," the function of which is to raise the circuit voltage by 0·5 volt per cell. This booster consists

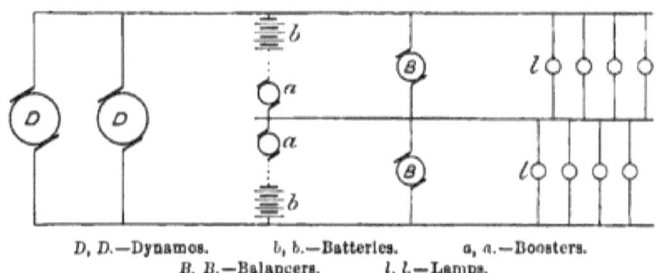

D, D.—Dynamos.   b, b.—Batteries.   a, a.—Boosters.
B, B.—Balancers.   l, l.—Lamps.

FIG. 94.—Arrangements for Three-wire Distribution.

of an electric motor coupled to a dynamo. The dynamo armature is so wound that it can pass without undue heating the charging current for the battery, and when running it adds an extra electromotive force in series with that of the main dynamos, and forces the charging current into the cells. When the cells are charged the booster is removed or stopped, and the cells can then deliver up current to the circuit after the main dynamos have been stopped. The arrangements usual in a three-wire continuous current station are shown diagrammatically in Fig. 94.

The two batteries act as regulators. If more lamps are switched on on one side of the three-wire system than on the other, the battery on that side gives out the necessary extra current whilst the main dynamos supply that which is taken by the remaining equal number of lamps on both sides of the middle wire.

Thus, suppose there are at any moment 10,000 8 c.p. lamps on one side, and 9,000 8 c.p. lamps switched on on the other, the main dynamo provides the current for the 9,000 lamps on each side of the middle wire, and the battery on one side furnishes the current for the odd 1,000 lamps on its own side.

Instead of the battery, balancing dynamos are sometimes employed. These are small dynamos, either driven independently, or else having their shafts coupled together. In the latter case each machine operates either as generator or motor according as the electromotive force of the machine is greater or less than the fall of potential across the two mains between which it is coupled. If the number of lamps on one side of the three-wire system exceeds that on the other, then one of the two coupled dynamos which is on the same side acts as a generator, and the other acts as a motor. Power is, therefore, taken from the more lightly loaded side and delivered to the other, and the equilibrium is restored, so that the current in the middle or neutral wire is kept nearly at zero.

In cases where the distances to be covered by the feeders are on an average greater than a mile and a-half or two miles, the high-pressure continuous current system is often employed. In this system the generating machines are continuous current dynamos generating at 1,000 or 2,000 volts pressure. The current from them is taken by high-pressure feeders to certain places called transformer stations. In these stations there is a machine (or machines) called a continuous current transformer. This appliance is a motor and dynamo combined in one. In some cases there are two armatures, each having its own field magnets. The one armature takes current at 1,000 to 2,000 volts pressure and acts as a motor, the other armature acts as a dynamo and generates current at 100 or 200 volts pressure. The low-pressure current is distributed by a two-wire low-pressure distribution system of mains used in private lighting.

In certain types of continuous current transformer the two armature windings are placed together, but insulated on one iron core, and there is but one field magnet. Each armature winding has, however, its own commutator. Very ingenious automatic switches have been designed which enable the engineer at the generating station to start and control any of the motor-generators in the distributing stations.

Any discussion on electric lamps and their use would be incomplete which did not include a brief reference to the subject of the measurement of electric energy by meter, as applied in the case of public electric supply. The complete solution of the commercial problem of distributing electric energy for use and sale is bound up with the practical one of the production of a satisfactory electric house meter.

Electric meters are broadly divided into two classes, called respectively *energy meters* and *quantity meters*. The first class measure electric energy passed through them to any apparatus or lamps, the second measure the total quantity of electricity which has passed through the meter. In so far as the supply companies or corporations are compelled to furnish electric current at a constant pressure, it follows that, for all practical purposes, the measurement of electrical quantity is equally as good as the measurement of electric energy, in those cases where the power taken up is measured by the product of the values of current and the voltage at

ELECTRIC DISTRIBUTION. 251

which it is supplied. In the case of alternating-current circuits, having on them motors or transformers, or arc lamps, it is necessary to employ energy meters, because in

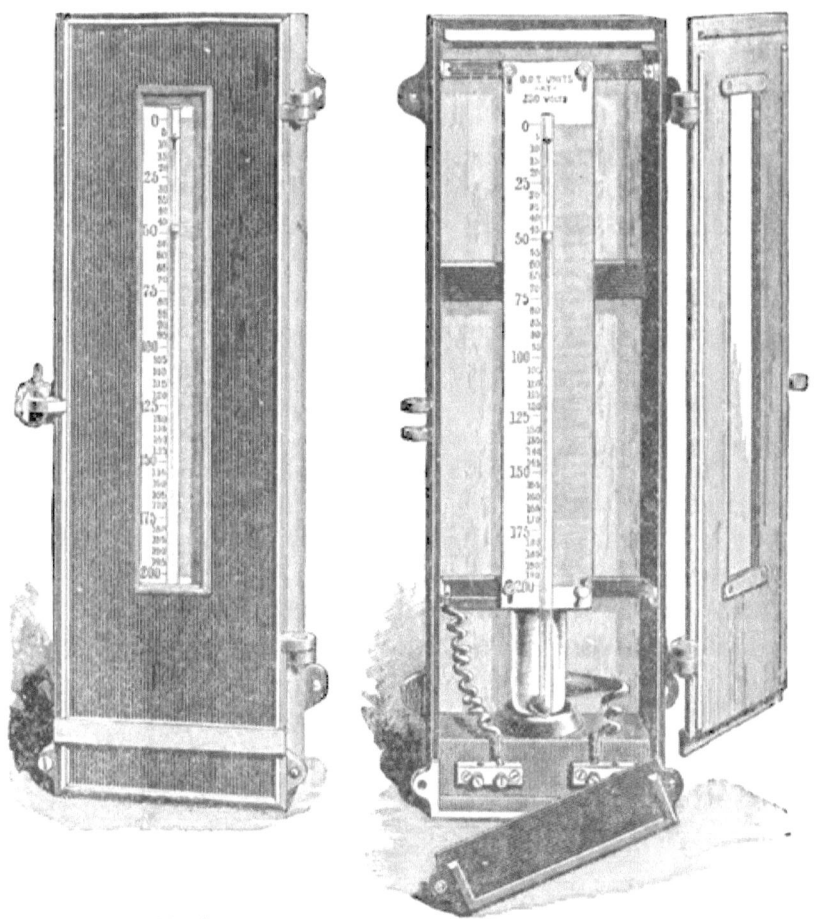

Closed.   Open.
Fig. 95.—Bastian's Electrolytic Meter.

those cases the power taken up is not always strictly measured by the numerical product of the current and the voltage.

A brief description may in the first place be given of some types of much used energy and quantity meters. One of the simplest forms of quantity meter, applicable, however, only to continuous current circuits, is Bastian's electrolytic meter. In this meter (*see* Fig. 95) the current entering the consumer's premises is made to pass through a glass vessel containing dilute sulphuric acid. The current is led into and out of the liquid by means of platinum plates, called electrodes. When the current passes through the liquid it chemically decomposes it, liberating gaseous oxygen and hydrogen, which escape. The amount of liquid is thereby reduced, and by means of a gauge glass the quantity of liquid left in the vessel can be ascertained at a glance. Every ampere-hour of electric quantity passed through the electrolyte causes the destruction of a certain volume of the liquid, and hence the gauge glass can be graduated to show at once the ampere-hours which have passed through the meter. To prevent evaporation from diminishing the volume of the acid, the surface is covered with a shallow layer of paraffin oil. This meter has the great advantage of extreme simplicity of construction and, therefore, of cheapness. It has the disadvantage that it requires filling up from time to time with water, and that it is not, without some complications, adapted for use with very large currents.

A type of ampere-hour meter in extensive use is one in which the principle of the electric motor is involved, and is hence called a motor meter. Of this form are the meters of Chamberlain and Hookham, Ferranti and Perry.

In the Chamberlain and Hookham meter, the rotating portion consists of two copper discs attached to a vertical spindle, and arranged so as to rotate in a magnetic field which is parallel to the axis of the discs. The current to be measured passes in at the centre of one disc and out at the edge, being so conducted by immersing the disc in mercury. The current passing radially through the disc thus placed in

ELECTRIC DISTRIBUTION. 253

a magnetic field at right angles to its plane, causes it to revolve. The associated disc also is rotated in the field, and is retarded by the so-called magnetic friction due to eddy electric currents set up in its mass. Hence a retarding force proportional to the speed is produced.

In the diagram (Fig. 96), M M are permanent magnets which have their poles in connection with iron pole pieces

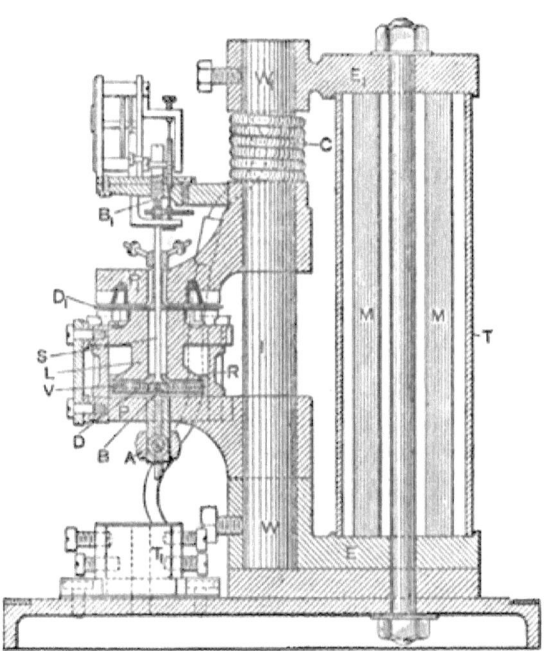

Fig. 96.—Chamberlain and Hookham Electric Meter.

P P_1. There is an intermediate iron piece, L, between the poles. This is held down on a copper ring, and forms a mercury chamber. In this chamber one of the copper discs revolves. The current passing through the disc causes it to rotate with a speed proportional to the current strength. The disc system is, however, retarded by the magnetic friction with a force proportional to the speed of rotation. Hence the speed is proportional to the current, and the number of revolutions in a given time to the total quantity of electricity which has in that time passed through the meter.

This meter, and others of a similar kind, have the advantage of very low internal resistance, and therefore, very small internal consumption of power. They measure

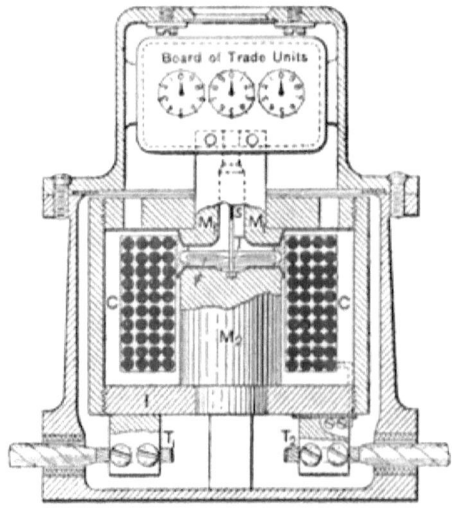

Fig. 97.—Ferranti Electric Meter.

directly the quantity of electricity passed through them, but it is indicated on the dials as Board of Trade units used, on the assumption that the circuit pressure is constant and equal to the certified standard electric pressure for the supply station.

Another ampere-hour meter of the same type, extensively used, is that of Ferranti (see Fig. 97). In this meter also a layer of mercury is contained in a thin cavity over an electromagnet. The current passing into the house or building using it is made to flow through the coils of the electromagnet, and through the mercury from the centre to the circumference. There are thus radial electric currents in the disc of mercury and a magnetic field perpendicular to its surface. Under these circumstances, the disc of mercury is set in revolution. The movement of the mercury is resisted by the friction of it against the sides of the cavity in which it is contained. The driving force is thus nearly proportional to the square of the current strength, and the retarding force to the square of the speed of revolution. Hence it follows that the number of revolutions in a given time is proportional to the quantity of electricity which has passed through the meter in that time. The revolutions of the mercury are counted by means of a delicate vane immersed in the mercury, which in turn drives a counting mechanism. The meter is adjusted so as to show on the dials the number of Board of Trade units of electrical energy which have passed through it, on the assumption that it is supplied at 100 volts. This meter is very compact. It is only, however, applicable in the case of continuous currents.

As a good example of an energy meter we may select for description the watt-hour meter of Elihu Thomson. This meter (*see* Fig. 98) is in reality a small dynamo, without iron either in its field magnets or armature. The field magnets are coils through which the current to be measured flows. The armature is a coil fixed as a shunt across the mains. Hence the magnetic field due to the current in the field coils is proportional to that current. The armature current is proportional to the voltage of the circuit. Accordingly the force driving the armature is proportional to the product of the values of the current and the voltage— that is, to the power given to the consumer. The motion of the armature is retarded by placing on its axis a copper disc, which moves between fixed permanent magnets. This disc has, then, eddy currents set up in it, and experiences a retarding force proportional to its speed. Accordingly the driving force is proportional to the product of the current and voltage of the main circuit, and the retarding force is proportional to the speed. Hence, when the armature has attained a steady speed, the number of revolutions it makes

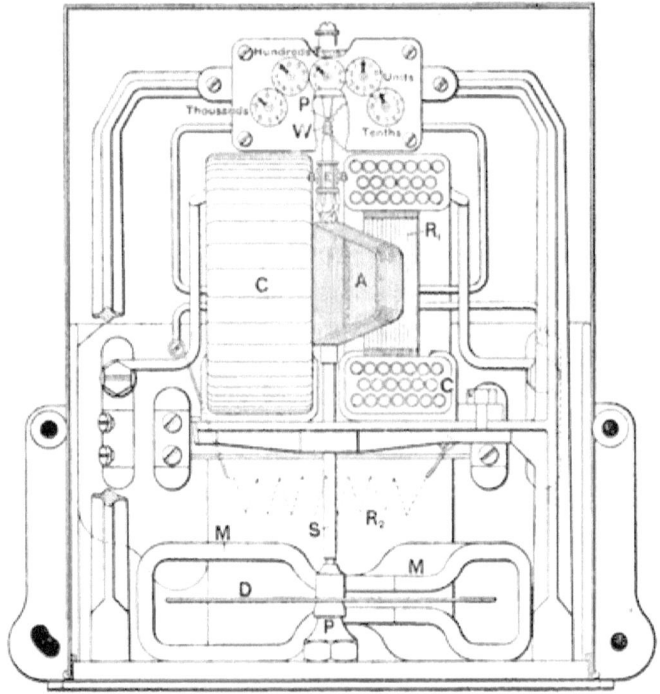

FIG. 98.—Elihu Thomson Electric Meter.

in any given time is proportional to the electrical energy which has passed through it. The armature shaft is attached to a counting mechanism, which shows on dials the energy in Board of Trade units which have passed through the meter. This meter is capable of being used either on alternating current or continuous current circuits.

In addition to the above-described motor and electrolytic meters, there is a large class of meters called integrating meters. In these appliances one portion of the apparatus consists of an ampere meter, or electric power meter. The remaining portion of the instrument consists of a clock recording time, and an integrating apparatus, which practically multiplies together, at short and regular intervals, the value of the current or power passing through the meter and the value of the time interval, and adds up the results, presenting the total as a time integral in the form of a record of ampere-hours or watt-hours on a series of recording dials. These meters are generally more complicated in construction than the motor meters. A good example of an integrating meter is the recently introduced Johnson and Phillips meter, of which a view is seen in Fig. 99.* The meter consists of an electrically-driven clock, actuating a counting mechanism through a gear whose ratio is controlled by the current or power passing to the lamps. The clock part is kept going by an ingenious arrangement which imparts a little push to the pendulum whenever the amplitude of its vibration falls below a certain value. The ammeter part of the instrument consists of a coil of wire, through which the house current passes. Into this is drawn a soft iron plunger consisting of a small bundle of iron wire, and the magnetic forces suck this plunger into the coil against the attractive forces of two non-magnetic springs. A needle in connection with the plunger indicates on a scale the current passing. The integrating part of the instrument consists of a revolving crank driven by the clock, which makes two revolutions a minute. Each revolution depresses a lever, and, by a ratchet and pawl arrangement, advances the first wheel of the counting dials. The extent to which the pawl is advanced each revolution is governed by the deflection of the ammeter needle. Hence the number of

---

* For further details see *The Electrician*, August 18, 1899.

revolutions of the counting dials in a given time is proportional to the time and to the deflection of the ammeter needle—that

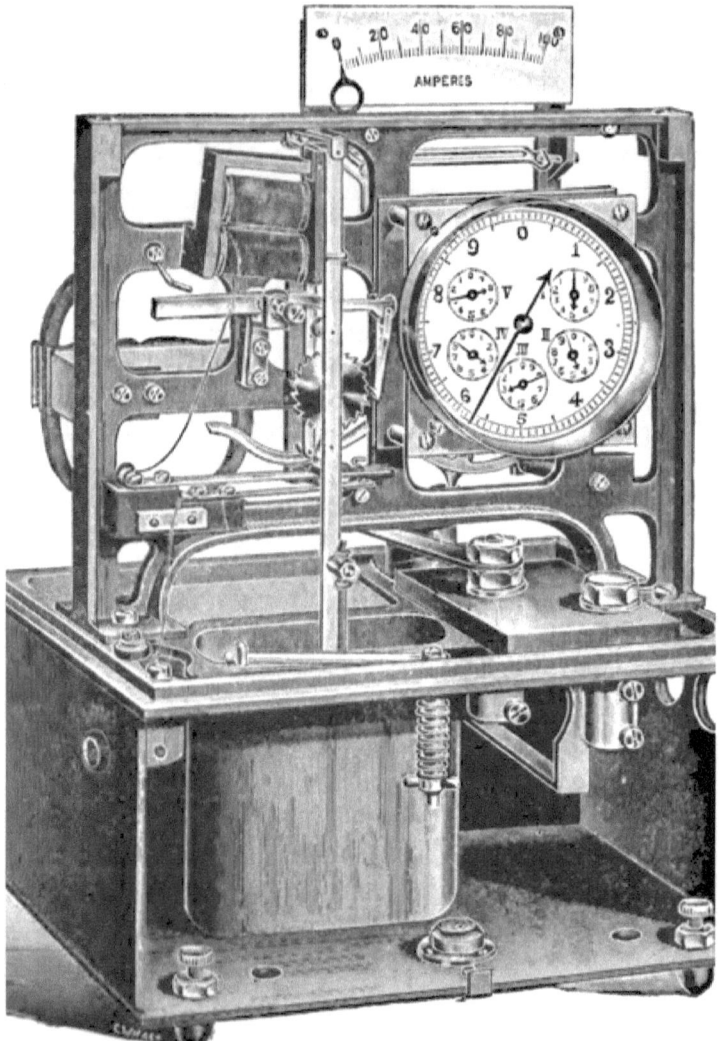

Fig. 99.—Front View of the Johnson and Phillips Meter.

is, to their product, or to the total quantity of electricity which has passed.

Considerable discussion has taken place amongst station engineers as to the relative advantages of quantity and energy meters. The former have the advantage in simplicity of construction, the latter in accuracy of reading under all conditions of varying voltage.

One great advantage which electric meters possess over gas meters is that the former are so easily checked for accuracy in position. If a householder having continuous current laid on desires to check his meter he should insert in series with the house meter an accurate ammeter, and across the terminals of the meter an accurate voltmeter. Then turn on a certain number of lamps in the house for one or more hours and carefully note the voltmeter reading and the ammeter readings. If these quantities vary during the time of observation, the mean value should be taken, the product of these readings gives the power in watts given to the lamps, and the product of watts and number of hours run, divided by 1,000, gives the energy in Board of Trade units taken up. This should agree with the difference between the house meter readings at the beginning and end of the run. The same process can be applied in the case of an alternating current system provided it is used only for supplying current to incandescent lamps. If motors are used on the circuit then it would be necessary to measure the power taken up with a properly designed alternating current wattmeter.

Quite recently meters have been designed called prepayment meters, which are set in operation by inserting a coin into a slot, and then deliver to the consumer a fixed amount of electric energy thus prepaid for.

One of the best known of these prepayment meters is that designed by Messrs. Long and Schattner. This meter is an electrolytic meter, and consists of a copper cell containing a solution of sulphate of copper, in which hangs a copper plate suspended from a lever A (*see* Figs. 100 and 101). At the end of this lever are two pins, H_1 H_2, which dip into mercury cups and close the main circuit. The lever also carries two buckets, K and E. The lever is so balanced that it just tips up and opens the circuit. If a coin (say one shilling) is then placed in the bucket K, it over-balances the lever and closes the circuit. The customer can now light up his lamps. The current to these lamps, however, partly passes through the solution of sulphate of copper, and takes copper off the

FIG. 100.—Long-Schattner Prepayment Meter.

suspended plate. When the weight of this is reduced by the weight of one shilling the lever A tilts up and opens the circuit. The adjustments are so made that the shilling put into the bucket just pays for the electric energy which has then passed. Another shilling's worth may then be obtained and so on. When the collector comes round to examine the meter and remove the coins, he places in the other bucket equivalent brass weights so that the meter goes on measuring out as before until the copper plate is entirely dissolved away. The whole meter is enclosed in a sealed case provided with a slot and guiding tube for receiving the coins.

ELECTRIC DISTRIBUTION. 261

In the commercial distribution and sale of electric energy it has been long known that the value of a consumer to the supply station does not depend so much upon the total amount of energy taken as upon the amount used per lamp per annum. A consumer who has one hundred lamps in his premises and uses them on an average one hour per day is not nearly so valuable to the supply station as one who has twenty lamps of the same kind and uses them on an average five hours a day. The former customer may in fact cause a loss to the suppliers. Hence of recent years a system has come into use (first suggested and worked out by Mr. A. Wright, chief engineer of the Brighton Corporation Electric

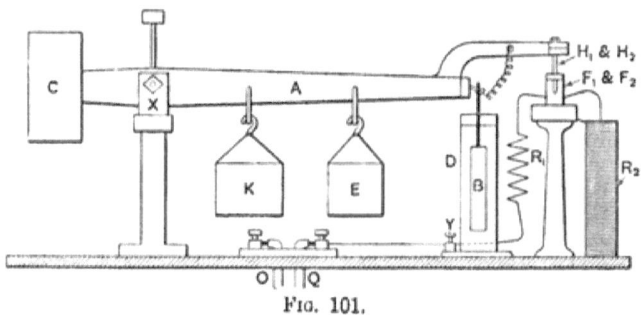
FIG. 101.

Supply Works), by which the average hours of use of the lamps in a consumer's premises is determined at the same time as his total consumption in electric energy units is measured. The charge per unit is then made on a sliding scale, so that the units for the first hour's use of the lamp each day are charged at a higher rate than the units for subsequent hours' use. The device for marking this measurement is called a Wright's maximum demand meter (see Fig. 102). It consists of a bent ∪ tube, having at one end a bulb surrounded by a coil of wire through which the customer's current passes. The other end of the tube has a bent-over trap bulb. The tube is partly filled with mercury. When the current passes through this coil it heats it and expands the air in the enclosed bulb. This forces some mercury over into the trap bulb, and when the current is stopped the amount by which the mercury in the left hand leg of the U

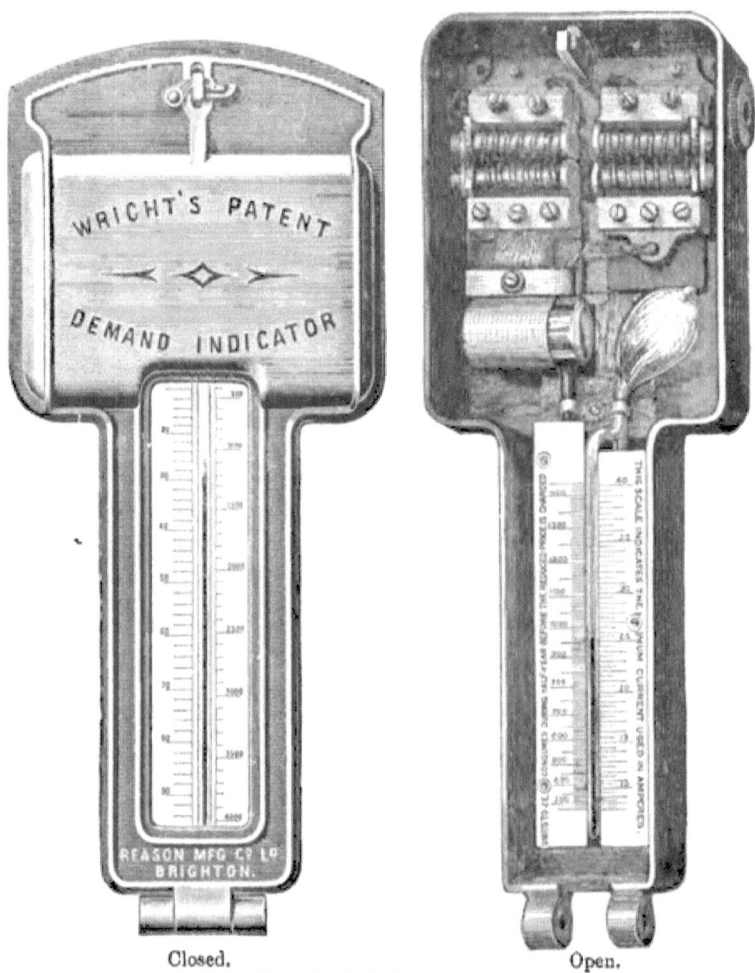

Closed.   Open.
Fig. 102.—Wright's Demand Indicator.

tube is depressed is a measure of the maximum current which has passed into the premises. The apparatus is re-set by tilting it, so as to make the mercury fall back into the tube.

The meter reader has, therefore, the power to discover not only the total units taken by the consumer, but also the maximum current or, which is the same thing, the maximum number of lamps he has ever had in use at one time. Hence a simple calculation shows the number of units per lamp, and, therefore, the average hours of use of each lamp.

Thus, suppose the maximum demand meter shows that a customer has taken a maximum current of 20 amperes, and his total units used in three months is 600 B.T.U. Reckoning his lamps, as 8 c.p. lamps and the service pressure as 100 volts, it is clear that the maximum power he has called for in watts is 2,000 watts, but his average power is 275 watts, since there are 2,184 hours in thirteen weeks, and $275 \times 2184 = 600,000$. Hence his average use of each lamp per day is $\frac{275}{2,000} \times 24 = 3 \cdot 3$ hours. He has used 600 units in 91 days. On the Wright system of charge the customer is not charged an equal amount for all units taken. He is charged, say, 7d. for the first two hours' use and $3\frac{1}{2}$d. for all subsequently used units. Hence, in the above case, the 600 units are charged as follows: 360 units at 7d., and 240 units at $3\frac{1}{2}$d. If his average use had been only two hours per day, it would all have been charged at 7d. If five hours per day the average cost per unit would only be 4·9d. Hence that customer gets an advantage who uses his lamps the longest number of hours, and not the one who takes most units.

The Wright system is now largely adopted, and has had a very beneficial effect in increasing the profits of the electric supply business.

For additional details on electric light fittings, systems of supply, and the construction of electric lamps, the reader must be referred to the other manuals which deal with each of these portions of the subject in a more complete manner than it has been possible to do in these short Lectures.

INDEX TO CONTENTS.

INDEX.

	PAGE
A Glance Backward Over Twenty Years	1
Absorption of Light	31
Ageing of Carbon Filament Lamps	88, 89, 133
Alternate Current Transformer	212
Alternator	218
Alternating Current Arc	189
Alternating Current	161
Ampere	9
——— The Definition of	10
Ampere-meter	69
Ampere turns	197
Amplitude	28
Annulment of Edison Effect	109
Arc, The Electric	151
Arc, Crater Temperature of	155
Arc Discharge	146
——— between Various Materials	166
Arc Lamps	191
——— Arrangement of	190
——— Carbons	179
——— Enclosed	185
——— Hand Regulated	149
——— Inverted	183
——— Mechanism	177
Arc, Appearance of the Electric	153
Armoured Cable	220
Artistic Electric Lighting	102
Bamboo Filament	56
Battery, Secondary	244
Bernstein Glow Lamp	58
Blackening of Incandescent Lamps	74
Board of Trade Unit	90

	PAGE
Board of Trade Electrical Inquiry of 1889	3
Booster	248
Bottomley's Experiments on Radiation	23
Brightness	33
——— Comparison of	34
Brush Discharge	140
Brussels, Electric Lighting Station at	241
Bunsen Photometer	42
Cable, Armoured	220
——— Concentric	220
Candle-Foot	38
Candle, Parliamentary, Standard	38
Candle-Power	69
——— Decrease of, in Incandescent Lamps	85, 86
——— of Incandescent Lamp, variation with Current	70
Carbon, Advantages of, for the Production of Electric Arcs	166
——— Allotropic Forms of	52
——— Boiling Point of	167
——— Deposit of, in Glow Lamps	75
——— Filaments	54, 55
——— ——— Properties of	31
——— ——— Ageing of	133
——— ——— Lamps made with	126
——— ——— Preparation of	54
——— ——— Process of Production	127
——— ——— Treated	130
——— Molecule, Negative Charge of	111

INDEX.

	PAGE
Horse-power, Definition of	20
Hydraulic Gradient	12
——— Power, Transmission of from Tivoli	228
Illuminating Power, Unit of	40
Illumination, Standard of	38
——— Unit of	40
Incandescent Lamp, Evolution of	52
——— Invention of	52
Incandescent Electric Lamps, Best Arrangement of, for Illuminating Purposes	98, 99
Incandescent Lamp	51, 54
——— Blackening of	74, 75
——— Different Forms of	56, 57, 58
Incandescent Lamps, High-Voltage	124
Incandescent Lighting of Interiors	100, 101
——— Artistic	99
——— Problem of	52
Intrinsic Brilliancy of Electric Arc Crater	173
——— of Glow Lamp Filament	175
——— of Sun	173
Invention of the Edison Glow Lamp	52
Inverted Arc Lamp	183
Iron Ring, Divided, with two Circuits	204
——— with Primary and Secondary Circuits	202
Joule, Mr.	18
——— The Definition of	20
Joule's Law	18
Kelvin Voltmeter	14
——— Multicellular Voltmeter	14
Kilowatt, Definition of the	21
Kinetic Theory of Gases	63
Lamp Efficiencies	126
Lane-Fox, Mr.	52
Life, Average, of Glow Lamps	135
Lightning Protector	157
Light, Monochromatic	31
——— Reflection of	31
——— Velocity of	28
——— Wave Lengths of Various Colours	29
Lines of Force	201
Load Diagram	223
——— Factor	95
Low-Pressure System	234
Luminosity	33

	PAGE
Luminosity of Various Coloured Surfaces	33
Luminous Efficiency	26
——— ——— of Glow Lamps	83
——— ——— of Incandescent Lamps	83
Magnetic Circuit	196
——— Circuits with Air Gaps	207
——— Effect of Current	195
——— Field, Delineation of	199
——— Field of an Electric Current	195
——— Field of Conductor	195, 196
——— Field of Magnetised Iron Rings	206
——— Field of Spiral Current	196
——— Force	203
——— Induction	203
——— Reluctivity	204
Magneto-motive Force	203
Magnetisation of Iron Circuits	205, 206
Manufacture of Incandescent Lamps	54
Marked Volts	68
Mean Free Path of Molecules	63
Mengarini Voltmeter	80
Mengarini's Self-recording Voltmeter	80
Mercury Pump, Form of	65
Metals, Non-Arcing	156
Methven Gas Standard	47
Micro-Glow Lamp	67
Molecular Electric Charge	111
——— Physics of Glow Lamp	106
——— Shadows in Glow Lamp	75
Moonlight, Intensity of	39
Multicellular Voltmeter	14
Multiple-Filament Glow Lamp	61
Nernst Lamp	136
Oersted, Prof. H. C.	7
——— H. C., his Discovery in 1820	7
Ohm, Dr. G. S.	9, 17
Ohm's Law	17
Parallel Arc Lamps	191
Parchmentised Thread	55
Pentane Standard, Vernon Harcourt.	47
Photometer, Bunsen	43
——— Grease Spot	44
——— Ritchie	42
——— Rumford	42
Photometric Comparisons	35
Photometry, Scientific Basis of	37
——— of Glow Lamps	40
——— of Various-Coloured Lights	46

INDEX.

	PAGE
Planté Secondary Battery	245
Platinum Standard of Light	47
Potential, Electric	11
Prepayment Meter	259
Pressure, Electric	11
———— Fall of, in Gas and Water Pipes	12
Primary Circuit	201
Principles of Photometry	36
Progress of Electric Lighting in Ten Years	4
Pump, Mercury	65
Purkinje's Phenomenon	36
Radiation at Different Temperatures	25
———— Luminous and Non-Luminous	26, 27
———— Gradual Production of, at Different Temperatures	24
Reflecting Power of Various Sources	104
Reluctivity	204
Resistance, Electric	16
Retrospect of Electric Lighting over Twenty Years	1
Ritchie's Photometer	40
———— Wedge	40
Rome, Electric Lighting at	225
Rumford's Shadow Photometer	42
Sawyer and Man	52
Secondary Battery	244
———— Cell	247
———— Circuit	205
Series Arc Lamps	190
———— or Parallel Lighting	191
Shades for Electric Arc Lamps	187
Sir Humphry Davy	146
Sir W. H. Preece	106
Smashing Point	86
Sources of Light, Qualities of	30
Spark Discharge	139
Sparking Distance	143
———— Pressure	144
Spectra, Similar	48
———— of Various Lights	49
Spectro-Photometer	49
Spectrum of a Source of Light	49
Spiral Electric Current, Field of	196
Standard, Candle	38
———— of Light	37
Stefan's Law	178
St. Pancras Electric Lighting Station	236
Summary of Facts concerning the Edison Effect on Glow Lamps	119
Street Arc Lamps	182
———— Arrangements of	191

	PAGE
Sun, Efficiency of	172
———— Intrinsic Brightness of	173
———— Temperature of	173
Sunlight, Comparison with other Luminous Sources	49
———— Intensity of	39
Table of Colours and Wave-Lengths of Light	29
———— of Reflecting Powers of Various Surfaces	104
———— showing Variation of Candle Power with Voltage of Incandescent Lamps	90
Temperatures, Important	25
———— Scale of	25
Terms, Fundamental, Meaning of	6
The Ampere, Definition of	10
The Electric Arc	151
The Joule	20
Three-Wire Distribution	221, 223
Three-wire Electric Lighting Station, Arrangement of Dynamos in	248
Tivoli, Electric Lighting Station at	227
Tivoli-Rome Transmission	226
———— Transformer House	231
Transformer, Construction of	211
———— Principle of	212
———— Sub-Centres	220
———— Systems	221
Transmission of Electric Power	210
Two Hundred Volt Service	124
Turbines at Tivoli	229
Untreated Carbon Filament Lamps	130
United Kingdom, Electric Lighting in	4
Units, Electric	9
Vacuum	54
———— Conductivity of	118
———— Tube	142
Variation of Candle-Power with Voltage	71
Variation of Candle-Power with Watts per Candle-Power	73
Variation of Candle-Power with Wattage in a Glow Lamp	73
Variation of Voltage of Public Electric Supply	77
Various Systems of Electric Distribution	219
Vernon Harcourt Pentane Standard	47
Violle Standard	47
Volt, Definition of the	16

INDEX.

	PAGE
Volta	16
Voltmeter	14, 68
———— Holden	78
———— Kelvin Multicellular	14
———— Mengarini	80
———— Self-recording	80
———— Self-recording (Mengarini)	80
Watt, Definition of the	20
Water Power	209
Water Pressure and Electric Pressure	12, 13, 210
Watts per Candle-Power	69
Wave Lengths	28
———— of Light	29
Welsbach Incandescent Lamp	137
Willans Engine	218
Wimshurst Electric Machine	13
Wiring, Electric	94
———— Inspection of	94
Work	20
Wright's Demand Indicator	262
Wright's System of Charge	263
Wurts' Discovery of Non-Arcing Metals	156
Wurts Lightning Protector	157

GEORGE TUCKER, PRINTER, SALISBURY COURT, FLEET STREET, LONDON.

Books for Electricians, Electrical Engineers, and Electrical Students. *Feb., 1900.*

THE Publisher of "THE ELECTRICIAN" has compiled the following List of Books on Electrical and Allied Subjects, all or any of which can be obtained direct from "THE ELECTRICIAN" PRINTING AND PUBLISHING COMPANY, LIMITED, Salisbury Court, Fleet Street, London, England. Telegrams and Cablegrams, "Electrician Newspaper London." Telephone No. **949** Holborn.

Any of the Works mentioned in this List will be forwarded to any address in the United Kingdom, on receipt of the quoted price. Five per cent. must be added for postage on foreign orders.

Copies of any known Work on the Theory or Practice of Electricity, or relating to any particular application of Electricity, or of an Electrical or Engineering character (either British or Foreign), can be ordered through this Company. Inquiries respecting the Latest and Best Books on Electrical Subjects will be answered *by post*, Free of Charge.

The Company supplies Books and Publications of every description at closest market prices, and is prepared to quote for Complete Library Outfits for Public or Private Libraries, especially in connection with the requirements of Technical Training Institutions.

Daily, Weekly, or Monthly parcels of Miscellaneous Books and Publications, for London, the Provinces, or Abroad, made up on the best possible terms.

Occasionally, slightly soiled copies of Electrical and other Books are on hand to be disposed of at reduced rates. Lists forwarded on application.

Cheques and Postal Orders to be made payable to "The Electrician Printing and Publishing Company, Limited," and to be crossed "Coutts and Co."

"The Electrician" Series of Standard Works.

[All fully Illustrated, and with very Complete Indices.]

	PAGE		PAGE
"The Electrician" Electrical Trades' Directory and Handbook	3	Carbon Making for all Electrical Purposes (*1900*)	15
Electromagnetic Theory (*Vols. I. and II.*)	4	Signalling Across Space Without Wires (*New and Enlarged Edition*)	13
Magnetic Induction in Iron and other Metals (*New Edition, 1900*)	4	Electrical Laboratory Notes and Forms (*Original and Cheaper Editions*)	12
Electric Lamps and Electric Lighting (*New Edition, 1900*)	5	Practical Notes for Electrical Students	13
Electrical Engineering Formulæ (*New Ed., 1900*)	5	Electrical Testing for Telegraph Engineers	13
The Alternate-Current Transformer in Theory and Practice (*Vols. I. and II.*)	6,7	Drum Armatures and Commutators	13
		Centenary of the Electric Current	13
Students' Guide to Submarine Cable Testing (*New Edition, 1899*)	7	Bibliography of X-ray Literature and Research	13
"The Electrician" Primers (*Vols. I. and II.*)	8	The Incandescent Lamp and its Manufacture	14
A Digest of the Law of Electric Lighting, Electric Traction, Telegraphs, Telephones, &c. (*1900*)	8	The Manufacture of Electric Light Carbons	14
		The Steam Engine Indicator, and Indicator Diagrams	14
Motive Power and Gearing	9	The Electric Arc	14
Electric Motive Power (*New Edition, 1899*)	9	Secondary Batteries, their Construction & Use	14
Localisation of Faults in Electric Light Mains	10	Primary Batteries, their Construction and Use	14
The Potentiometer and its Adjuncts	10	Commercial Telephony	14
Submarine Cable Laying and Repairing	10	"The Electrician" Wireman's Pocket Book	14
The Art of Electrolytic Separation of Metals	11	Fine Art Reproductions, &c.	15
Electro-Chemistry	11	Steel Plate Portraits	15

NEW VOLUMES in preparation.—*See page 14.*

Electrical and General List.

"The Electrician"	32	Telegraphy and Telephony	24
Newspapers and Periodicals	2	Medical Electricity	25
The "Engineering Magazine"	2	Electrical Instruments, Bells, &c.	26
International Wire Tables	4	Röntgen Ray Literature	20
Armature Windings of Electric Machines	15	Photography, &c.	26
Electricity and Magnetism	16	Dictionaries, Directories, &c.	27
Electric Lighting and Transmission of Power	19	International Telegraph Convention	15
		Sheet Tables of Expenditure, &c., of Electricity Stations	15
Wiring and Electric Light Fitting	23	Guide to the Correction of Errors in Code and other Telegrams	8
Electro-Chemistry and Electro-Metallurgy	25	Miscellaneous	

PAPERS AND PERIODICALS.

"The Electrician" Printing and Publishing Company (Limited) forward, post free, on receipt of the *annual subscription*, any of the following Papers and Periodicals:—

AMERICAN.

AMERICAN ELECTRICIAN. (Monthly.) Ann. sub., 8s. 6d.
AMERICAN JOURNAL OF SCIENCE AND ARTS. (Monthly.) Ann sub., 26s. 5d.
ELECTRICAL ENGINEERING (Semi-monthly.) Ann. sub., 6s. 6d.
ELECTRICAL REVIEW. (Weekly.) Ann. sub., £1. 1s.
ELECTRICAL WORLD and ENGINEER. (Weekly.) Ann. sub., £1. 5s.; single copies 8d.; post free, 10d.
ELECTRICITY. (Weekly.) Ann. sub., 19s.
ENGINEERING MAGAZINE. (Monthly.) Ann. sub., 12s. 6d.; abroad 14s.; single copies, 1s.; post free, 1s. 3d.
JOURNAL OF PHYSICAL CHEMISTRY. (Ten times a year.) Ann. sub., 10s.
PHYSICAL REVIEW. (Monthly.) Ann. sub., 21s.
RAILWAY REVIEW. (Weekly.) Ann. Sub., £1. 1s.
CASSIER'S MAGAZINE. Single copies 1s. 3d.
ENGINEERING NEWS. (Weekly.) Ann. sub., 30s.
SCIENTIFIC AMERICAN. (Weekly.) Ann. sub., 17s.
SCIENTIFIC AMERICAN SUPPLEMENT. (Weekly.) Ann. sub., £1. 5s.
STREET RAILWAY JOURNAL. (Monthly.) Ann. sub., £1. 5s.
STREET RAILWAY REVIEW. (Monthly.) Ann. sub., 16s. 6d.
WESTERN ELECTRICIAN. (Weekly.) Ann. sub. £1. 1s.

CONTINENTAL.

ANNALEN DER PHYSIK UND CHEMIE (Germany). (Twelve parts per annum.) 36s.
ANNALES DE CHIMIE ET DE PHYSIQUE (France). (Monthly.) Ann. sub., 29s.
BULLETIN INTERNATIONAL DE L'ÉLECTRICITÉ (France). (Weekly.) Ann. sub., 8s.
COMPTES RENDUS, Organ of the Paris Academy of Sciences (France). (Weekly.) £1. 7s.
L'ECLAIRAGE ELECTRIQUE (France). (Weekly.) Ann. sub., £2. 8s.
L'ELECTRICIEN (France). (Weekly.) Ann. sub., £1.
L'ELECTRICIEN (Russia). (Monthly.) Ann. sub., 7 roubles.
ELEKTRITCHESTWO (Russia). (Fortnightly.) Ann. sub., 12 roubles.
ELECTROCHEMISCHE ZEITSCHRIFT (Germany). (Monthly.) Ann. sub., 18s. 6d.
ELEKTROTECHNISCHE ZEITSCHRIFT (Germany). (Weekly.) Ann. sub., £1. 5s.; single copies, 6d. post free, 8d.
ELEKTROTECHNISCHER ANZEIGER (Germany). (Semi-weekly.) Ann. sub., 10s.
ELEKTROTECHNISCHER RUNDSCHAU (Germany). (Bi-weekly.) Ann. sub., 12s.
ELEKTROTECHNISCHES ECHO (Germany). (Weekly.) Ann. sub., 16s.
L'ELETTRICISTA (Italy). (Monthly.) Ann. sub., 12s.
L'ELETTRICITA (Italy). (Weekly.) Ann. sub., 12s.
L'ENERGIE ELECTRIQUE (Italy). (Bi-weekly.) Ann. sub., 15s.
IL NUOVO CIMENTO (Italy). (Monthly.) 10s.
L'INDUSTRIE (France). (Weekly.) Ann. sub., £1.
L'INDUSTRIE ELECTRIQUE (France). (Semi-monthly.) Ann. sub., £1. 1s.
JOURNAL DE PHYSIQUE THÉORIQUE ET APPLIQUÉE (France). (Monthly.) 12s. 6d.
JOURNAL TÉLÉGRAPHIQUE (Switzerland). (Monthly.) Ann. sub., 5s.
KRAFT UND LICHT (Germany). (Weekly.) Ann. sub., 10s.
REVUE PRATIQUE DE L'ELECTRICITÉ (France). (Bi-weekly.) Ann. sub., 16s.
ZEITSCHRIFT FUR ELECTROCHEMIE (Germany). (Fortnightly.) 16s.
ZEITSCHRIFT FUR ELEKTROTECHNIK (Austria). (Weekly.) Ann. sub., 16s. 6d.
ZEITSCHRIFT FUR INSTRUMENTENKUNDE (Germany). (Monthly.) Ann. sub., £1.
ZEITSCHRIFT FUR PHYSIKALISCHE CHEMIE (Germany). (Three vols. per annum.) 15s

---

Annual Subscription, 12s. 6d. *post free (U.K.); European Continent, &c.,* 14s.

"THE ENGINEERING MAGAZINE,"
(European Edition.)

A High-class Monthly Periodical having an extensive sale. Each number presents special articles by distinguished authorities, together with a very valuable Review and Index to current Engineering literature.

Sales Agent—**GEO. TUCKER, 1, 2 & 3, Salisbury Court, Fleet Street, London, England.**

1, 2 and 3, Salisbury Court, Fleet Street, London, E.C.

The Electrician" Printing and Publishing Co., Ltd.,

PUBLISHED IN JANUARY OF EACH YEAR. ROYAL 8vo., about 1,250 pages.
Subscription Price, **6s.**; *post free* **6s. 9d.**, *abroad* **7s. 6d.** *After date of Publication,* **12s. 6d.**, *post free* **13s. 3d.** (*Colonies,* **14s.**, *post free; other countries,* **15s.**)

"THE ELECTRICIAN"
ELECTRICAL TRADES' DIRECTORY
AND HANDBOOK.
THE CHEAPEST, LARGEST, AND ONLY RELIABLE DIRECTORY PUBLISHED FOR THE ELECTRICAL AND KINDRED TRADES.
(THE BIG BLUE BOOK.)

This Directory and Handbook is well known throughout the Electrical, Engineering, and Allied Trades in all parts of the World, and no expense is spared to make the work *really reliable*. New names and corrections to existing entries in the DIRECTORIAL portion of the book are received up to the middle of January, no charge being made for inserting names, addresses, and occupations. Extra insertions are made in the Classified Trades' Section at very low rates. In addition to more than 600 pages of purely Directorial matter, the HANDBOOK portion contains a mass of interesting information, the great part of which cannot be obtained elsewhere, and in the Biographical Division appear about 300 sketches of the lives of well-known men in the Electrical World, with many excellent Portraits.

Following are some of the **Permanent Features** of this indispensable Reference Book, which has been aptly designated "the combined Kelly and Whitaker of the Electrical Trades":—

A DIRECTORY OF THE PROFESSIONS AND TRADES CONNECTED WITH ELECTRICITY AND ITS APPLICATIONS:—
BRITISH DIVISION.
COLONIAL DIVISION.
CONTINENTAL DIVISION.   } Alphabetically Arranged and Classified into
ASIATIC & AFRICAN DIVISION.   Trades and Professions.
CENTRAL & S. AMERICAN DIVISION.
UNITED STATES DIVISION.

Summary of Electrical Events for the preceding year. Obituary Notices for the preceding year.
PATENTS, DESIGNS, AND TRADE MARKS.—(*Specially written.*)
Principal British Electrical Patents expiring in the near future. [&c., &c.—(*Special.*)
DIGEST OF THE LAW OF ELECTRIC LIGHTING, of ELECTRIC TRACTION, TELEGRAPHS, TELEPHONES.
Revised Board of Trade Regulations Concerning the Supply of Electricity.
German High-Pressure Regulations.
Board of Trade Regulations as to Railways in Metal-Lined Tunnels
London County Council Regulations as to Theatre Lighting, Overhead Wires, Electric Meter Testing, &c.
Rules and Regulations for the Prevention of Fire Risks arising from Electric Lighting: British and Foreign—
Lloyd's Suggestions for the Use of Electric Light on Board Vessels, [(*Special.*)
Electric Lighting and Electric Traction Notices for the year.
Provisional Orders and Licenses granted by the Board of Trade in the preceding year
Light (Electric) Railway applications granted in the preceding year.
New Companies Registered and Companies Wound Up in the preceding year.
CENTRAL LIGHTING STATIONS OF THE UNITED KINGDOM.—(*Specially-Compiled Folding Sheet.*)
COLOURED SKETCH MAP OF PROVINCIAL SUPPLY STATIONS, with System of Supply, &c. (*Special.*
A SHEET TABLE GIVING TECHNICAL PARTICULARS OF THE ELECTRIC RAILWAYS AND TRAMWAYS IN THE UNITED KINGDOM.—(*Specially-Compiled Folding Sheet.*)
Exports of Electrical Apparatus and Material. Foreign Weights and Measures, with English Equivalents.
Colonial and Foreign Import Duties on Electrical Machinery and Apparatus.—(*Specially prepared.*)
Table of Systems of Charging for Electric Current, "Easy" Wiring, "Free" Lamps, Motors, &c.
Wire Gauge Tables. Resistance and Weight Tables. Comparative Price of Electricity and Gas.
Tables relating to Water Power, Hydraulic Heads, Rope Gearing, &c. Sinking Fund Tables.
Electric Traction Tables and Data. Carrying Capacity of Resistance Materials.
Notes on Illumination. Table and Curve of Hours of Lighting. Post Office Regulations. Electric Heating Table
TELEGRAPHIC COMMUNICATION between Great Britain, Colonies, and Foreign Countries. } (*Specially*
TELEGRAPH TARIFFS from the United Kingdom to all parts of the World. } *prepared.*)
The Submarine Cables of the World. The Land Lines of the World. The World's Cable Fleet.
Telephone Statistics, including Post Office Telephone Trunk Line Regulations, Forms of Telephone Licence for Local Authorities and Companies, British Telephone Developments, &c
BRITISH AND FOREIGN GOVERNMENTS AND THE CHIEF OFFICIALS OF ELECTRICAL ENGINEERING

"THE ELECTRICIAN" SERIES.

Vol. I. SECOND ISSUE. 466 pages, price 12s. 6d., post free, 13s.
Vol II. 568 pages, price 12s. 6d., post free ; abroad, 13s.

ELECTROMAGNETIC THEORY.

By OLIVER HEAVISIDE.

EXTRACT FROM PREFACE TO VOL. I.

This work is something approaching a connected treatise on electrical theory, though without the strict formality usually associated with a treatise. The following are some of the leading points in this volume. The first chapter is introductory. The second consists of an outline scheme of the fundamentals of electromagnetic theory from the Faraday-Maxwell point of view, with some small modifications and extensions upon Maxwell's equations. The third chapter is devoted to vector algebra and analysis, in the form used by me in former papers. The fourth chapter is devoted to the theory of plane electromagnetic waves, and, being mainly descriptive, may perhaps be read with profit by many who are unable to tackle the mathematical theory comprehensively. I have included in the present volume the application of the theory (in duplex form) to straight wires, and also an account of the effects of self-induction and leakage, which are of some significance in present practice as well as in possible future developments.

EXTRACT FROM PREFACE TO VOL. II.

From one point of view this volume consists essentially of a detailed development of the mathematical theory of the propagation of plane electromagnetic waves in conducting dielectrics, according to Maxwell's theory, somewhat extended. From another point of view, it is the development of the theory of the propagation of waves along wires. But on account of the important applications, ranging from Atlantic telegraphy, through ordinary telegraphy and telephony, to Hertzian waves along wires, the author has usually preferred to express results in terms of the concrete voltage and current, rather than the specific electric and magnetic forces belonging to a single tube of flux of energy. . . . The theory of the latest kind of so called wireless telegraphy (Lodge, Marconi, &c.) has been somewhat anticipated, since the waves sent up the vertical wire are hemispherical, with their equatorial bases on the ground or sea, which they run along in expanding. (See § 60, Vol. I. ; also § 393 in this volume.) The author's old predictions relating to skin conduction, and to the possibilities of long-distance telephony have been abundantly verified in advancing practice: and his old predictions relating to the behaviour of approximately distortionless circuits have also received fair support in the quantitative observation of Hertzian waves along wires.

SECOND ISSUE (1900). 370 pages, 150 illustrations. Price 10s. 6d., post free.

MAGNETIC INDUCTION IN IRON AND OTHER METALS.

By J. A. EWING, M.A., B.Sc.,

Professor of Mechanism and Applied Mechanics in the University of Cambridge.

SYNOPSIS OF CONTENTS.

After an introductory chapter, which attempts to explain the fundamental ideas and the terminology, an account is given of the methods which are usually employed to measure the magnetic quality of metals. Examples are then quoted, showing the results of such measurements for various specimens of iron, steel, nickel and cobalt. A chapter on Magnetic Hysteresis follows, and then the distinctive features of induction by very weak and by very strong magnetic forces are separately described, with further description of experimental methods, and with additional numerical results. The influence of Temperature and the influence of Stress are next discussed. The conception of the Magnetic Circuit is then explained, and some account is given of experiments which are best elucidated by making use of this essentially modern method of treatment. The book contains a chapter on the Molecular Theory of Magnetic Induction, and the opportunity is taken to refer to a number of miscellaneous experimental facts on which the molecular theory has an evident bearing. The concluding chapter deals with the important subject of Practical Magnetic Testing.

Strongly bound in Cloth, price 5s., post free.

COMPREHENSIVE INTERNATIONAL WIRE TABLES

FOR ELECTRIC CONDUCTORS.

By W. S. BOULT.

Giving very Full Particulars of **Wires** and **Cables** ranging from 2·26in. to ·001in. diameter, and including Board of Trade, Standard, Birmingham, Brown and Sharpe, and Metric Gauges to the number of 469, arranged in consecutive order according to cross-section, so that when any conductor is sought the nearest sizes in the other gauges can at once be found. Any No. of any one of the gauges named, Single Wire or Cable, can be readily ascertained. Full particulars are given of Cross-section, Gauge, Weight and Resistance in English, American and Continental Units, one Table supplying the place of the various independent Tables which have hitherto been published. Great pains have been taken throughout these Tables to ensure accuracy.

"THE ELECTRICIAN" SERIES—*continued*.

NOW READY.—NEW AND CHEAPER EDITION (1900).
Very fully illustrated, handsomely bound, on good paper, price 6s.

ELECTRIC LAMPS AND ELECTRIC LIGHTING.

By PROF. J. A. FLEMING, M.A., D.Sc., F.R.S., M.R.I.,
Professor of Electrical Engineering in University College, London.

SYNOPSIS OF CONTENTS.

I.—A Retrospect over 25 Years—Present Condition of Electric Lighting—Chief Properties of an Electric Current—Names of Electrical Units—Chemical Power of an Electric Current—Hydraulic Analogies—Electric Pressure—Fall of Electric Pressure Down a Conductor—Ohm's Law—Joule's Law—Units of Work and Power—The Watt as a Unit of Power—Incandescence of a Platinum Wire—Spectroscopic Examination of a Heated Wire—Visible and Invisible Radiation—Luminous Efficiency—Radiation from Bodies at Various Temperatures—Efficiency of Various Sources of Light—The Glow Lamp and Arc Lamp as Illuminants—Colours and Wavelengths of Rays of Light—Similar and Dissimilar Sources of Light—Colour-Distinguishing Power—Causes of Colour—Comparison of Brightness and Colour—Principles of Photometry—Limitations Due to the Eye—Luminosity and Candle Power—Standards of Light—Standards of Illumination—The Candle Foot—Comparison of Sunlight and Moonlight—Comparison of Lights—Ritchie's Wedge—Rumford and Bunsen Photometers—Comparison of Lights of Different Colours—Spectro Photometers—Results of Investigations.

II.—The Evolution of the Incandescent Lamp—The Nature of the Problem—Necessary Conditions—Allotropic Forms of Carbon—The Modern Glow Lamp—Processes for the Manufacture of the Filament—Edison-Swan Lamps—The Expansion of Carbon when Heated—Various Forms of Glow Lamps—Velocity of Molecules of Gases—Kinetic Theory of Gases—Processes for the Production of High Vacua—Necessity for a Vacuum—Mean Free Path of Gaseous Molecules—Voltage, Current and Candle Power of Lamps—Watts per Candle Power—Characteristic Curves of Lamps—Life of Glow Lamps—Molecular Shadows—Blackening of Glow Lamps—Self-Recording Voltmeters—Necessity for Constant Pressure of Supply—Changes Produced in Lamps by Age—Smashing Point—Efficiency of Glow Lamps—Statistics of Age—Variation of Candle Power with Varying Voltage—Cost of Incandescent Lighting—Useful Life of Lamps—Importance of Careful "Wiring"—Average Energy Consumption of Lamps in Various Places—Load Factors—Methods of Glow Lamp Illumination for Production of Best Effects—Artistic Electric Lighting—Molecular Physics of the Glow Lamp—High Voltage Lamps—Varieties of Carbon—Densities and Resistances—Deterioration of Carbon Lamps—High Efficiency Lamps—Recently Suggested Improvements.

III.—Forms of Electric Discharge—Vacuum Tubes—Sparking Distance—The Electric Arc—The Optical Projection of the Arc—The Arc a Flexible Conductor—The High Temperature of the Arc—Non-Arcing Metals—Lightning Protectors—The Distribution of Light from the Arc—Continuous and Alternating Current Arcs—Voltage Required to Produce an Arc—The Physical Actions in the Arc—The Changes in the Carbons—The Distribution of a Potential in the Arc—The Unilateral Conductivity of the Arc—The Temperature of the Crater—Comparison with Solar Temperature—The "Watts per Candle" of the Sun—Intrinsic Brightness and Dissipative Power of Heated Surfaces—Comparison of Glow Lamp, Arc Lamp and Sun in Respect of Brightness and Radiation—Arc Lamp Mechanism—Arc Lamp Carbons—The Hissing of Arc Lamps—The Applications of Arc Lamps—Inverted Arcs—Series and Parallel Arc Lighting—The Enclosed Arc Lamp.

IV.—The Generation and Distribution of Electric Current—The Magnetic Action of an Electric Current—The Magnetic Field of a Spiral Current—The Induction of Electric Currents—The Peculiar Magnetic Property of Iron—Iron and Air Magnetic Circuits—The Typical Cases of an Iron Conduit With and Without Air Gaps—The Prototypical Forms of Dynamo and Transformer—The Transformation of Electrical Energy—Hydraulic Illustrations—The Mechanical Analogue of a Transformer—The Mode of Construction of an Alternate Current Transformer—The Fundamental Principle of all Dynamo Electric Machines—Alternating and Direct Current Dynamos—Alternating Current Systems of Electrical Distribution—Description of the Electric Lighting Station in Rome—The Tivoli-Rome Electric Transmission—Views of the Tivoli Station—Continuous Current Systems—The Three-Wire System—Description of St. Pancras Vestry Electric Lighting Station—Liverpool, Glasgow and Brussels Electric Lighting Stations—Direct Driven Dynamos—Alternating and Continuous Current Systems Contrasted—Secondary Batteries—House and Maximum Demand Meters.

NEW EDITION READY EARLY IN 1900.
800 pages, specially bound bookwise to lie open, **7s. 6d.** *nett, post free* **7s. 9d.** *(abroad,* **8s.,** *post free).*

A POCKET-BOOK OF
ELECTRICAL ENGINEERING FORMULÆ, &c.

By W. GEIPEL AND H. KILGOUR.

With the extension of all branches of Electrical Engineering (and particularly the heavier branches), the need of a publication of the Pocket-Book style dealing practically therewith increases; for while there are many such books referring to Mechanical Engineering, and several dealing almost exclusively with the lighter branches of electrical work, none of these suffice for the purposes of the numerous body of Electrical Engineers engaged in the application of electricity to Lighting, Traction, Transmission of Power, Metallurgy, and Chemical Manufacturing. It is to supply this real want that this most comprehensive book has been prepared.

Compiled to some extent on the lines of other pocket-books, the rules and formulæ in general use among Electricians and Electrical Engineers all over the world have been supplemented by brief and, it is hoped, clear descriptions of the various subjects treated, as well as by concise articles and hints on the construction and management of various plant and machinery.

No pains have been spared in compiling the various sections to bring the book thoroughly up to date; and while much original matter is given, that which is not original has been carefully selected, and, where necessary, corrected. Where authorities differ, as far as practicable a mean has been taken, the differing formulæ being quoted for guidance.

Specimen Pages, &c., sent post free on application.

NEW EDITION—Almost entirely Rewritten, and brought up to date.
More than 600 pages and 215 Illustrations, 12s. 6d. post free; abroad, 13s.

THE ALTERNATE CURRENT TRANSFORMER
IN THEORY AND PRACTICE.
By J. A. FLEMING, M.A., D.Sc., F.R.S., M.R.I., &c.,
Professor of Electrical Engineering at University College, London.

Since the first edition of this Treatise was published, the study of the properties and applications of alternating electric currents has made enormous progress. The author has, accordingly, rewritten the greater part of the chapters, and availed himself of various criticisms, with the desire of removing mistakes and remedying defects of treatment. In the hope that this will be found to render the book still useful to the increasing numbers of those who are practically engaged in alternating-current work, he has sought, as far as possible, to avoid academic methods and keep in touch with the necessities of the student who has to deal with the subject not as a basis for mathematical gymnastics but with the object of acquiring practically useful knowledge.

SYNOPSIS OF CONTENTS OF VOL. I.

CHAPTER I.—Introductory.—Faraday's Electrical Researches—Faraday's Theories—Magneto-Electric Induction—Views of Maxwell, Helmholtz and Kelvin—Action at a Distance—The Electro-Magnetic Medium—Joseph Henry's Investigations.

CHAPTER II.—Electro-Magnetic Induction.—Magnetic Force and Magnetic Fields—Magnetic Force in Conductors—Typical Cases—Magneto-Motive Force and Magnetic Induction—Flux of Force—Magnetic Permeability—Lines of Induction—Faraday's Law of Induction—Electromotive Force Due to the Change of Induction—The Magnetic Circuit—Magnetic Resistance—Lines of Induction of Closed and Open Magnetic Circuits—Fundamental Relations Between Magnetic Force and Magnetic Induction—Intensity of Magnetisation—Magnetic Moment—Lines and Tubes of Magnetic Induction—Curves of Magnetisation—Permeability Curves—Determination of Permeability—Magnetic Hysteresis—Hysteresis Curves—H.B. Diagram—Effect of Temperature on Hysteresis—Electromotive Force of Induction—Graphical Representations—Electromotive Force of Self-Induction.

CHAPTER III.—The Theory of Simple Periodic Currents.—Variable and Steady Flow—Current and Electromotive Force Curves—Fourier's Theorem—Mechanical Harmonograph—Mathematical Sketch of Fourier's Theorem—Practical Application of Fourier's Theorem to the Harmonic Analysis of a Periodic Curve—Simple Periodic Currents and Electromotive Forces—Description of a Simple Periodic Curve—The Mean Value of the Ordinate of a Sine Curve—The Square Root of the Mean of the Squares of the Ordinates of a Simple Periodic Curve—Derived Curves—Inductance and Inductive Circuits; Inductance, Resistance and Capacity of Circuits; Inductive and Non-Inductive Circuits—Faraday's and Henry's Experiments on Self-Induction—Edison's and Maxwell's Arrangement for Exhibiting Inductance of Circuits—Electro-Magnetic Momentum—Electrotonic State and Electromagnetic Energy—Co-efficient of Mutual Induction; Energy of Two Circuits—The Unit of Inductance—Value of the Self-Induction in Henry's for Various Instruments—Current Growth in Inductive Circuits—Analogy of Current and Velocity Change; Fundamental Equations for Current Growth in Inductive Circuits—Equation for the Establishment of a Steady Current in Inductive Circuits—Time Constant of an Inductive Circuit—Logarithmic Curves; Instantaneous Value of Simple Periodic Current—Solution of Current Equation—Impedance of Inductive Circuit—Relations of Impressed Electromotive Force, Current and Impedance—Geometrical Illustrations—Impressed and Effective Electromotive Force—Clock Diagram of Electromotive Forces in Inductive Circuit—Triangle Representing Resistance, Impedance and Reactance of Circuit—The Mean Value of the Power of a Periodic Current—Geometrical Theorem—Power Curves for Inductive and Non-Inductive Circuits—Experimental Measurement of Periodic Currents and Electromotive Forces—Mean Square Value—Method of Measuring the True Mean Power Given to an Inductive Circuit—Theory of the Wattmeter—Divided Circuits—Important Trigonometrical Lemma—Impedance of Branched Circuits—Wattmeter Measurement of Periodic Power—Mutual Induction of Two Circuits of Constant Inductance—The Flow of Simple Periodic Currents into a Condenser—Time Constant of a Condenser—Charging a Condenser through a Resistance—Condenser equation—Annulment of Inductance by Capacity—Representation of Periodic Currents by Polar Diagrams—Initial Conditions on Starting Current Flow in Inductive Circuits—Complex Periodic Functions—Apparent and True Power given to Inductive Circuits—Power Factor.

CHAPTER IV.—Mutual and Self-Induction.—The Researches of Joseph Henry—Experiments with Coils and Bobbins; Discovery of Self-Induction—Mutual Induction—Induction at a Distance—Induction between Telephone Circuits—Induction over Great Distances—Induced Currents of Higher Orders—Inductive Effects by Transient Electric Currents—Magnetic Screening—Direction of Induced Currents—Various Qualities of an Induced Current—Elementary Theory of Mutual Induction of Two Circuits—Theory of Induction Coil with Non-Magnetic Core—Comparison of Theory and Experiment—Duration of Induced Currents—Magnetic Screening Action of Good-Conducting Masses—Faraday's and Henry's Experiments—Willoughby Smith's Investigations on Magnetic Screening—Dove's Experiments and Henry's Views on same—The Reaction of the Secondary Currents on the Primary Circuit in the Case of an Induction Coil—Induction Balance and Sonometer—Transmission of Alternating Currents through Conductors—Prof. Hughes' Experiments—Lord Rayleigh's Researches—Flow of Current through Conductor—Surface Flow of Alternating Currents—Increased Resistance of Conductors for Alternating Currents of High Frequency—Limiting Size of Conductors for Conveyance of Alternating Currents—Stephan's Analogies—Electro-Magnetic Repulsion—Elihu Thomson's Experiments—Electro-Magnetic Rotations.

CHAPTER V.—Dynamical Theory of Induction.—Electro-Magnetic Theory—Faraday's Conception of an Electro-Magnetic Medium—Maxwell's Suggestion—The Luminiferous Ether—Maxwell's Theory of Electric Displacement—Electric Elasticity of the Medium—Displacement and Conduction Currents—Electromotive Intensity—Displacement Currents and Displacement Waves—Theory of Molecular Vortices—Mechanical Analogy—Comparison of Theory and Experiment—Maxwell's Law Connecting Dielectric Constant and Refractivity of a Dielectric—Tables of Comparison—Velocity of Propagation of Electro-Magnetic Disturbances—Values of "v"—Vector Potential—Electrical Oscillation—Charge and Discharge of Leyden Jar—The Function of the Condenser in an Induction Coil—Impulsive Discharges and Relation of Inductance Thereto—Impulsive Impedance—Hertz's Researches—Experimental Determination of the Velocity of Electro-Magnetic Waves.

CHAPTER VI.—The Induction Coil and Transformer.—General Description of the Action of the Transformer or Induction Coil—The Delineation of Periodic Curves of Current and Electromotive Force—Curve Tracers—Transformer Diagrams—Curves of Electromotive Force—Current and Induction in Cases of Various Transformers Taken off Various Alternators—Open-Circuit Current of Transformers—Symmetry of Transformer Curves—Harmonic Analysis of Transformer Curves—Power and Hysteresis Curves—Hysteresis Curves of Various Transformers—Efficiency of Transformers—Efficiency Curves of Transformers—Tables of Efficiencies—Current Diagrams of a Transformer—Tables of Complete Tests—The Power Factor—Open and Closed Circuit Transformers—Magnetic Leakage and Secondary Drop—Various Causes of Secondary Drop—Determination of Magetic Leakage—Investigations of the Author and Dr. Roessler—Form Factor and Amplitude Factor of a Periodic Curve—General Analytical Theory of the Transformer and Induction Coil.

"The Electrician" Printing and Publishing Co., Ltd.,

"THE ELECTRICIAN" SERIES—*continued.*

THIRD ISSUE. *More than 600 pages and over 300 illustrations. Price* **12s. 6d.**, *post free; abroad,* **13s.**

THE ALTERNATE CURRENT TRANSFORMER
IN THEORY AND PRACTICE.

By J. A. FLEMING, M.A., D.Sc., F.R.S., M.R.I., &c.,

Professor of Electrical Engineering in University College London.

SYNOPSIS OF CONTENTS OF VOL. 2.

CHAP. I.—Historical Development of Induction Coil and Transformer.
The Evolution of the Induction Coil—Page's Researches—Callan's Induction Apparatus—Sturgeon's Induction Coil—Bachhoffner's Researches—Callan's Further Researches—Callan's Great Induction Coil—Page's Induction Coil—Abbot's Coil - Automatic Contact Breakers—Ruhmkorff's Coils — Poggendorff's Experiments — Stöhrer's, Hearder's, Ritchie's Induction Apparatus—Grove's Experiments—Apps' Large Induction Coils—Jablochkoff's Patent—Fuller's Transformer—Early Pioneers—Goulard and Gibbs—Zipernowsky's Transformers—Improvements of Rankin Kennedy, Hopkinson, Ferranti, and others—The Modern Transformer since 1885.

CHAP. II.—Distribution of Electrical Energy by Transformers.
Detailed Descriptions of Large Alternate-Current Electric Stations using Transformers in Italy, England, and United States—Descriptions of the Systems of Zipernowsky-Déri-Bláthy, Westinghouse, Thomson-Houston, Mordey, Lowrie Hall, Ferranti, and others — Plans, Sections, and Details of Central Stations using Transformers—Illustrations of Alternators and Transformers in Practical Use in all the chief British, Continental, and American Transformer Stations.

CHAP. III.—Alternate-Current Electric Stations.
General Design of Alternating-Current Stations, Engines, Dynamos, Boilers—Proper Choice of Units—Water Power—Parallel Working of Alternators—Underground Conductors—Various Systems—Concentric Cables—Capacity Effects dependent on Use of Concentric Cables—Phenomena of Ferranti Tubular Mains—Safety Devices—Regulation of Pressure—Choice of Frequency—Methods of Transformer Distribution—Sub-Stations—Automatic Switches.

CHAP. IV.—The Construction and Action of Transformers.
Transformer Indicator Diagrams—Ryan's Curves—Curves of Current—Electromotive Force and Induction-Analysis of Transformer Diagrams—Predetermination of Eddy Current and Hysteresis Loss in Iron Cores—Calculation and Design of Transformers—Practical Predetermination of Constants—Practical Construction of Transformers—Experimental Tests of Transformers—Measurement of Efficiency of Transformers—Calometric Dynamometer and Wattmeter Methods—Reduction of Results.

CHAP. V.—Further Practical Application of Transformers.
Electrical Welding and Heating Transformers for producing Large Currents of Low Electromotive Force—Theory of Electric Welding—Other Practical Applications—Conclusion.

NEW EDITION.—JUST PUBLISHED.
Fully Illustrated. Price **6s.** *net, post free; abroad* **6s. 3d.**
STUDENTS' GUIDE TO
SUBMARINE CABLE TESTING.
By H. K. C. FISHER and J. C. H. DARBY.

The authors of this book have, for some years past, been engaged in the practical work of Submarine Cable Testing in the Eastern Extension Telegraph Company's service, and have embodied their experience in a Guide for the use of those in the Telegraph Service who desire to qualify themselves for the examinations which the Cable Companies have recently instituted. To those desirous of entering the Cable Service, Messrs. Fisher and Darby's book is indispensable, as it is now necessary for probationers to pass these examinations as part of the qualification for service.

1, 2, and 3, Salisbury Court, Fleet Street, London, F.C.

"THE ELECTRICIAN" SERIES—*continued.*

In Two Volumes, 2s. 6d. *each, post free* 2s. 9d. *each.   Single Primers,* 3d., *post free* 3½d.

"THE ELECTRICIAN" PRIMERS.

(FULLY ILLUSTRATED.)

A Series of Helpful Primers on Electrical Subjects for the use of those seeking a Knowledge of Electricity—Theoretical and Practical.

TABLE OF CONTENTS.

Volume I.—THEORY.

Primer No.
1. The Effects of an Electric Current.
2. Conductors and Insulators.
3. Ohm's Law.
4. Primary Batteries.
5. Arrangement of Batteries.
6. Electrolysis.
7. Secondary Batteries.
8. Lines of Force.
9. Magnets.
10. Electrical Units.
11. The Galvanometer.
12. Electrical Measuring Instruments.
13. The Wheatstone Bridge.
14. The Electrometer.
15. The Induction Coil.
16. Alternating Currents.
17. The Leyden Jar.
18. Influence Machines.
19. Lightning Protectors.
20. Thermopiles.

The object of "The Electrician" Primers is to briefly describe in simple and correct language the present state of electrical knowledge. Each Primer is short and complete in itself, and is devoted to the elucidation of some special point or the description of some special application. Theoretical discussion is as far as possible avoided, the principal facts being stated and made clear by reference to the uses to which they have been put. Both volumes are suited to readers having little previous acquaintance with the subject. The matter is brought up to date, and the illustrations refer to instruments and machinery in actual use at the present time. It is hoped that the Primers will be found of use wherever the want of a somewhat popularly written work on electricity and its industrial applications, published at a popular price, has been felt. Electricity Committees of Town Councils will find the Primers of great service. Artisans will find the Primers useful in enabling them to obtain clear notions of the essential principles underlying the apparatus of which they may be called upon to take charge.

Volume II.—PRACTICE.

Primer No.
21. The Electric Telegraph.
22. Automatic and Duplex Telegraphy.
23. The Laying and Repair of Submarine Cables.
24. Testing Submarine Cables.
25. The Telephone.
26. Dynamos.
27. Motors.
28. Transformers.
29. The Arc Lamp.
30. The Incandescent Lamp.
31. Underground Mains.
32. Electric Meters.
33. Electric Light Safety Devices.
34. Systems of Electric Distribution.
35. Electric Transmission of Energy.
36. Electric Traction.
37. Electro-Deposition.
38. Electric Welding.

New Edition in March, price 3s., *post free.*

A DIGEST OF THE
LAW OF ELECTRIC LIGHTING, ELECTRIC TRACTION, AND OTHER SUBJECTS.

By A. C. CURTIS-HAYWARD, B.A., M.I.E.E.

Being a full critical abstract of the Electric Lighting Acts, 1882 and 1889, of the Tramways Act, 1870, and of the various documents emanating from the Board of Trade dealing with Electric Lighting and Electric Traction, including the Rules as to the procedure in connection with applications to the Light Railway Commissioners for Orders under the Light Railways Act, 1896. The Digest treats first of the manner in which persons desirous of supplying electricity must set to work, and then of their rights and obligations after obtaining Parliamentary powers; and gives in a succinct form information of great value to Local Authorities, Electric Light Contractors, &c., up to date. The Board of Trade Regulations as to the Supply of Electrical Energy, the London County Council Regulations as to Overhead Wires, Theatre Lighting, &c., together with the Bye-laws enforced in pursuance of Part II. of the Public Health Acts Amendment Act, 1890, by the various Urban Sanitary Authorities are also given.

---

440 pp.   Fourth Edition.   Price 7s. 6d.

GUIDE TO THE
CORRECTION OF ERRORS IN CODE AND OTHER TELEGRAMS.

Containing about 70,000 examples, many of which are the outcome of actual experience. A large proportion of the presumed errors in telegrams may be explained by a reference to these pages. Invaluable to Telegraph and Cable Companies, and to Commercial Houses using Code

"**THE ELECTRICIAN" SERIES**—*continued.*

Price 12s. 6d., *post free; abroad,* 13s.

MOTIVE POWER AND GEARING
FOR ELECTRICAL MACHINERY:

A Treatise on the Theory and Practice of the Mechanical Equipment of Power Stations for Electric Supply, and for Electric Traction.

By E. TREMLETT CARTER, C.E., M.I.E.E., F.R.A.S., F.P.S. (Lond.), &c.

650 *pages,* 200 *Illustrations, Scale Drawings and Folding Plates, and over* 80 *Tables of Engineering Data.*

IN ONE VOLUME.

Part I.—Introductory. Part II.—The Steam Engine. Part III.—Gas and Oil Engines.
Part IV.—Water Power Plant. Part V.—Gearing. Part VI.—Types of Power Stations

This work presents to consulting engineers, contractors, central-station engineers, and engineering students the latest and most approved practice in the equipment and working of mechanical plant in electric-power generating stations. Every part of the work has been brought completely up to date; and especially in the matter of the costs of equipment and working the latest available information has been given. The treatise deals with Steam, Gas, Oil and Hydraulic Plant and Gearing; and it deals with these severally from the three standpoints of (1) Theory, (2) Practice and (3) Costs

"MOTIVE POWER AND GEARING FOR ELECTRICAL MACHINERY" is a handbook of modern electrical engineering practice in all par. f the world. It offers to the reader a means of comparing the central-station practice of the United Kingdom with that of America, the Colonies or other places abroad; and it enables him to study the scientific, economic and financial principles upon which the relative suitability of various forms of practice is based, and to apply these principles to the design or working of plant for any given kind of work, whether for electrical supply or for electric traction. It is a treatise which should be in the hands of every electrical engineer throughout the world, as it constitutes the only existing treatise on the Economics of Motive Power and Gearing for Electrical Machinery.

NEW EDITION (1900).

Over 400 pages, nearly 250 illustrations. Price **10s. 6d.**, *post free; abroad,* **11s.**

ELECTRIC MOTIVE POWER.

By ALBION T. SNELL, Assoc.M.Inst.C.E., M.I.E.E.

The rapid spread of electrical work in collieries, mines, and elsewhere has created a demand for a practical book on the subject of transmission of power. Though much had been written, there was no single work dealing with the question in a sufficiently comprehensive and yet practical manner to be of real use to the mechanical or mining engineer; either the treatment was adapted for specialists, or it was fragmentary, and power work was regarded as subservient to the question of lighting. The Author has felt the want of such a book in dealing with his clients and others, and in "ELECTRIC MOTIVE POWER" has endeavoured to supply it.

In the introduction the limiting conditions and essentials of a power plant are analysed, and in the subsequent chapters the power plant is treated synthetically. The dynamo, motor, line, and details are discussed both as to function and design. The various systems of transmitting and distributing power by continuous and alternate currents are fully enlarged upon, and much practical informatio gathered from actual experience, is distributed under the various divisions. The last two chapters deal exhaustively with the applications of electricity in mining work in Great Britain, the Continent and America, particularly with reference to collieries and coal-getting, and the results of the extensive experience gained in this field are embodied.

In general, the Author's aim has been to give a sound digest of the theory and practice of the electrical transmission of power, which will be of real use to the practical engineer, and to avoid controversial points which lie in the province of the specialist, and elementary proofs which properly belong to text-books on electricity and magnetism.

To meet the convenience of Continental readers and others, the Author has prepared in tabular form and in parallel columns the working equations used in this work in *inch-pound-minute* and *centimetre-gramme-second* units, so that they may be readily used in either system.

1, 2 and 3, Salisbury Court, Fleet Street, London, E.C

"THE ELECTRICIAN" SERIES—continued.

Price **5s.**, *post free; abroad,* **5s. 6d.** *180 pages and over 100 illustrations.*

THE LOCALISATION OF FAULTS IN ELECTRIC LIGHT MAINS.

By F. CHARLES RAPHAEL.

Although the localisation of faults in telegraph cables has been dealt with fully in several hand-books and pocket-books, the treatment of faulty electric light and power cables has never been discussed in an equally comprehensive manner. Beyond a few short articles which have appeared in the technical journals from time to time, nothing has been written on the subject. The conditions of the problems presented for solution are, however, very different in the two cases; faults in telegraph cables are seldom localised before their resistance has become low compared with the resistance of the cable itself, while in electric light work the contrary almost always obtains. This fact alone entirely changes the method of treatment required in the latter case, and it has been the author's endeavour, by dealing with the matter systematically, and as a separate subject, to adequately fill a gap which has hitherto existed in technical literature.

The various methods of insulation testing during working have been collected and discussed, as these tests may be considered to belong to the subject.

Price **6s.**, *post free; abroad* **6s. 6d.**

THE POTENTIOMETER AND ITS ADJUNCTS.

(A UNIVERSAL SYSTEM OF ELECTRICAL MEASUREMENT.)

By W. CLARK FISHER.

The extended use of the Potentiometer System of Electrical Measurement will, it is hoped, be sufficient excuse for the publication of this work, which, while dealing with the main instrument, its construction, use and capabilities, would necessarily be incomplete without similar treatment of the various apparatus which, as adjuncts, extend the range and usefulness of the whole system.

Electrical testing may be said to have passed through two stages. First, that which may be called the elementary, in which first principles were evolved; secondly, the adaptation of the same to the needs of the telegraph and cable engineer. But with the advent of electric lighting and other undertakings, such testing might be said to have passed into the third or practical and commercial stage, where large quantities have to be dealt with, and where the old order of things changeth.

The engineer or practical man demands that he shall be shown results quickly, plainly and accurately with a minimum of trouble, understanding, and consequently "Time," and on that account prefers—like all good mechanics—to have one good instrument, which, once understood and easily manipulated, can be used in a variety of ways to suit his needs. It is to this fact undoubtedly that the "Potentiometer" method of measurement owes its popularity. Its accuracy is rarely, if ever, impugned. Measurements made by it are universally accepted amongst engineers, and it might be well termed a "universal" instrument in "universal" use.

Over 400 pages and 200 specially drawn Illustrations. Price **12s. 6d.**, *post free.*

SUBMARINE CABLE-LAYING AND REPAIRING.

By H. D. WILKINSON, M.I.E.E., &c., &c.

This work describes the procedure on board ship when removing a fault or break in a submerged cable and the mechanical gear used in different vessels for this purpose; and considers the best and most recent practice as regards the electrical tests in use for the detection and localisation of faults, and the various difficulties that occur to the beginner. It gives a detailed technical summary of modern practice in Manufacturing, Laying, Testing and Repairing a Submarine Telegraph Cable. The testing section and details of boardship tests have been prepared with the object and hope of helping men in the cable services who are looking further into these branches. The description of the equipment of cable ships and the mechanical and electrical work carried on during the laying and repairing of a submarine cable will also prove to some not directly engaged in the profession, but nevertheless interested in the enterprise, a means of informing themselves as to the work which has to be done from the moment a new cable is projected until it is successfully laid and worked.

The Chapter on "*Testing*" *is especially valuable and up to date.*

"**THE ELECTRICIAN" SERIES**—*continued.*

Over 300 pages, 106 illustrations.   Price **10s. 6d.**, *post free.*

The ART of ELECTROLYTIC SEPARATION of METALS
(THEORETICAL AND PRACTICAL).
By GEORGE GORE, LL.D., F.R.S.

THE ONLY BOOK ON THIS IMPORTANT SUBJECT IN ANY LANGUAGE.

SYNOPSIS OF CONTENTS.

HISTORICAL SKETCH.
Discovery of Voltaic and Magneto-Electricity—First Application of Electrolysis to the Refining of Copper—List of Electrolytic Refineries.

THEORETICAL DIVISION.
Section A.: *Chief Electrical Facts and Principles of the Subject.*—Electric Polarity and Induction, Quantity, Capacity, Potential—Electromotive Force—Electric Current—Conduction and Insulation—Electric Conduction Resistance.
Section B.: *Chief Thermal Phenomena.*—Heat of Conduction Resistance—Thermal Units Symbols, and Formulæ.
Section C.: *Chief Chemical Facts and Principles of the Subject.*—Explanation of Chemical Terms—Symbols and Atomic Weights—Chemical Formulæ and Molecular Weights—Relation of Heat to Chemical Action.
Section D.: *Chief Facts of Chemico-Electric or Voltaic Action.*—Electrical Theory of Chemistry—Relation of Chemical Heat to Volta Motive Force—Volta-Electric Relations to Metals in Electrolytes—Voltaic Batteries—Relative Amounts of Voltaic Current produced by Different Metals.
Section E.: *Chief Facts of Electro-Chemical Action.*—Definition of Electrolysis—Arrangements for Producing Electrolysis—Modes of Preparing Solutions—Nomenclature—Physical Structure of Electro-Deposited Metals—Incidental Phenomena attending Electrolysis—Decomposability of Electrolytes—Electro-Chemical Equivalents of Substances—Consumption of Electric Energy in Electrolysis.
Section F.: *The Generation of Electric Currents by Dynamo Machines.*—Definition of a Dynamo and of a Magnetic Field—Electro-Magnetic Induction—Lines of Magnetic Force.

PRACTICAL DIVISION.
Section G. *Establishing and Working an Electrolytic Copper Refinery.*—Planning a Refinery—Kinds of Dynamos Employed—Choice and Care of Dynamo—The Depositing Room—The Vats—The Electrodes—The Main Conductors—Expenditure of Mechanical Power and Electric Energy—Cost of Electrolytic Refining.
Section H.: *Other Applications of Electrolysis in Separating and Refining Metals.*—Electrolytic Refining of Copper by other Methods—Extraction of Copper from Minerals and Mineral Waters—Electrolytic Refining of Silver Bullion and of Lead—Separation of Antimony, of Tin, of Aluminium, of Zinc, of Magnesium, of Sodium and Potassium, of Gold—Electrolytic Refining of Nickel—Electric Smelting.
Appendix.—Useful Tables and Data.

Second Edition, price **2s.**, *post free.*

ELECTRO-CHEMISTRY.
By GEORGE GORE, LL.D., F.R.S.

This book contains, in systematic order, the chief principles and facts of electro-chemistry and is intended to supply to the student of electro-plating and electro-metallurgy a scientific basis upon which to build the additional practical knowledge and experience of his trade. A scientific foundation, such as is here given of the art of electro-metallurgy is indispensable to the electro-depositor who wishes to excel in his calling, and should be studied previously to and simultaneously with practical working. As the study of electro-chemistry includes a knowledge not only of the conditions under which a given substance is electrolytically separated, but also of the electrolytic effect of a current on individual compounds, both are described, and the series of substances are treated in systematic order. *An indispensable book to Electro-Metallurgists.*

1, 2 and 3, Salisbury Court, Fleet Street, London, E.C.

"THE ELECTRICIAN" SERIES—*continued*.

Electrical Laboratory Notes & Forms.

ARRANGED AND PREPARED BY

Dr. J. A. FLEMING, M.A., F.R.S.

Professor of Electrical Engineering in University College, London.

These "Laboratory Notes and Forms" have been prepared to assist Teachers, Demonstrators and Students in Electrical Laboratories, and to enable the Teacher to economise time. They consist of a series of (about) Twenty Elementary and (about) Twenty Advanced Exercises in Practical Electrical Measurements and Testing. For each of these Exercises a four-page Report Sheet has been prepared, two pages of which are occupied with a condensed account of the theory and practical instructions for performing the particular Experiment, the other two pages being ruled up in lettered columns, to be filled in by the Student with the observed and calculated quantities. Where simple diagrams will assist the Student, these have been supplied. These Exercises are for the most part based on the methods in use in the Electrical Engineering Laboratories of University College, London; but they are perfectly general, and can be put into practice in any Electrical Laboratory.

Each Form is supplied either singly at **4d.** nett, or at **3s. 6d.** per dozen nett (assorted or otherwise as required); in sets of any three at **1s.** nett; or the set of (about) Twenty Elementary (or Advanced) Exercises can be obtained, price **5s. 6d.** nett. The complete set of Elementary and Advanced Exercises are price **10s. 6d.** nett, or in a handy Portfolio, **12s.** nett, or bound in strong cloth case, price **12s. 6d.** nett.

Spare Tabulated Sheets for Observations, price **1d.** each nett.

Strong Portfolios, price **1s. 6d.** each.

The very best quality foolscap sectional paper (16in. by 13in.) can be supplied, price **1s** per dozen sheets nett.

NOW READY.—Cheaper edition of "Electrical Laboratory Notes and Forms." These cheaper Forms have been prepared for the use of students and teachers at the Polytechnic and other science classes throughout the country. These new Forms, which differ only from the higher-priced set in being printed on smaller and cheaper paper and with less space for tabulated records, are issued at half the price of the original set.

ELEMENTARY SERIES.

1. The Exploration of Magnetic Fields.
2. The Magnetic Field of a Circular Current.
3. The Standardisation of a Tangent Galvanometer by the Water Voltameter
4. The Measurement of Electrical Resistance by the Divided Wire Bridge.
5. The Calibration of the Ballistic Galvanometer.
6. The Determination of Magnetic Field Strength.
7. Experiments with Standard Magnetic Fields
8. The Determination of the Magnetic Field in Air Gap of an Electro-magnet.
9. The Determination of Resistance with the Post Office Pattern Wheatstone Bridge.
10. The Determination of Potential Difference by the Potentiometer.
11. The Measurement of a Current by the Potentiometer.
12. A Complete Report on a Primary Battery.
13. The Standardisation of a Voltmeter by the Potentiometer.
14. The Photometric Examination of an Incandescent Lamp.
15. The Determination of the Absorptive Powers of Semi-Transparent Screens
16. The Determination of the Reflective Power of Various Surfaces.
17. The Determination of the Electrical Efficiency of an Electromotor by the Cradle Method.
18. The Determination of the Efficiency of an Electromotor by the Brake Method.
19. The Efficiency Test of a Combined Motor-Generator Plant.
20. Efficiency Test of a Gas Engine and Dynamo Plant.

ADVANCED SERIES.

21. The Determination of the Electrical Resistivity of a Sample of Metallic Wire.
22. The Measurement of Low Resistances by the Potentiometer.
23. The Measurement of Armature Resistances.
24. The Standardisation of an Ammeter by Copper Deposit.
25. The Standardisation of a Voltmeter by the Potentiometer.
26. The Standardisation of an Ammeter by the Potentiometer.
27. The Determination of the Magnetic Permeability of a Sample of Iron.
28. The Standardisation of a High Tension Voltmeter.
29. The Examination of an Alternate-Current Ammeter.
30. The Delineation of Alternating Current Curves.
31. The Efficiency Test of a Transformer.
32. The Efficiency Test of an Alternator.
33. The Photometric Examination of an Arc Lamp.
34. The Measurement of Insulation and High Resistance
35. The Complete Efficiency Test of a Secondary Battery.
36. The Calibration of Electric Meters.
37. The Delineation of Hysteresis Curves of Iron.
38. The Examination of a Sample of Iron for Magnetic Hysteresis Loss
39. The Determination of the Capacity of a Concentric Cable.
40. The Hopkinson Test of a Pair of Dynamos

"THE ELECTRICIAN" SERIES—continued.

320 pages, 155 illustrations. Price 6s. 6d., post free.

PRACTICAL NOTES FOR ELECTRICAL STUDENTS.
LAWS, UNITS, AND SIMPLE MEASURING INSTRUMENTS.
By A. E. KENNELLY and H. D. WILKINSON, M.I.E.E.

CONTENTS.

Chapter I.—Introductory.
,, II.—Batteries.
,, III.—Electromotive Force and Potential.
,, IV.—Resistance.
,, V.—Current.

Chapter VI.—Current Indicators.
,, VII.—Simple Tests with Indicators.
,, VIII.—Calibration of Current Indicators.
,, IX.—Magnetic Fields and their Measurement.

Now Ready.—*Very fully Illustrated. Price 10s. 6d. ; post free, 11s.*

Electrical Testing for Telegraph Engineers.
By J. ELTON YOUNG.

This book embodies up-to-date theory and practice in all that concerns everyday work of the Telegraph Engineer.

CONTENTS.

Chapter I.—Remarks on Testing Apparatus.
,, II.—Measurements of Current, Potential, and Battery Resistance.
,, III.—Natural and Fault Current.
,, IV.—Measurement of Conductor Resistance.
,, V.—Measurement of Insulation Resistance.
,, VI.—Corrections for Conduction and Insulation Tests.

Chapter VII.—Measurement of Inductive Capacity.
,, VIII.—Localisation of Disconnections.
,, IX.—Localisation of Earth and Contacts.
,, X.—Corrections of Localisation Tests.
,, XI.—Submarine Cable Testing during Manufacture, Laying and Working.
,, XII.—Submarine Cable Testing during Localisation and Repairs.

In the Appendices numerous tables and curves of interest to telegraph engineers are given.

Fully Illustrated. Price 7s. 6d., post free.

DRUM ARMATURES AND COMMUTATORS
(THEORY AND PRACTICE).
By F. MARTEN WEYMOUTH.

NOW READY.

British Association Evening Discourse at Dover, September 18, 1899.

THE CENTENARY OF THE ELECTRIC CURRENT
By Prof. J. A. FLEMING, F.R.S.   1799–1899.

With Illustrations of early apparatus and interesting Chronological Notes.

In neat paper covers 1/-, post free 1/3.   Bound cloth 2/-, post free.

Very fully Illustrated. Price 5s. nett ; post free 5s. 3d.

Wireless Telegraphy.
NEW and ENLARGED EDITION NOW READY.

SIGNALLING ACROSS SPACE WITHOUT WIRES.
By Dr. OLIVER J. LODGE, F.R.S.

The New Edition forms a complete Illustrated Treatise on Hertzian Wave Work. The Full Notes of the interesting Lecture delivered by the Author before the Royal Institution in June, 1894, form the first chapter of the book. The second chapter is devoted to the Application of Hertz waves and Coherer Signalling to Telegraphy, while Chapter 3 gives Details of other Telegraphic Developments. In Chapter 4 a history of the Coherer Principle is given, including Professor Hughes' Early Observations before Hertz or Branly, and the work of M. Branly. Chapters are also devoted to "Communications with respect to Coherer Phenomena on a Large Scale" the "Photo-Electric Researches of Drs. Elster and G. itel" and the Photo-Electric Researches of Prof. Righi.

Price 5s. post free ; abroad 5s. 3d.

The BIBLIOGRAPHY OF X-RAY LITERATURE and RESEARCH.

Being a carefully and accurately compiled Ready Reference Index to the Literature on the Subject of Röntgen or X-Rays.

Edited by CHARLES E. S. PHILLIPS.

With an Historical Retrospect and a Chapter, "Practical Hints," on X-Ray work by the Editor.

The book will be indispensable for the Reference Library, Public Libraries, Colleges, Technical Schools and Laboratories, and to all whose business or pleasure necessitates an up-to-date acquaintance with the progress of this branch of Physical Science.

"THE ELECTRICIAN" SERIES—*continued.*

NOW READY.
Fully Illustrated. Price **10s. 6d.** *Postage free.*

CARBON MAKING FOR ALL ELECTRICAL PURPOSES.
By FRANCIS JEHL.

This work gives a concise account of the process of making High Grade and other Carbon for Electric Lighting, Electrolytic, and all other electrical purposes.

CONTENTS.

- Chapter I.—Physical Properties of Carbon.
- " II.—Historical Notes.
- " III.—Facts concerning Carbon.
- " IV.—The Modern Process of Manufacturing Carbons.
- " V.—Hints to Carbon Manufacturers and Electric Light Engineers.
- " VI.—A "New" Raw Material.
- " VII.—Gas Generators.
- " VIII.—The Furnace.
- Chapter IX.—The Estimation of High Temperatures.
- " X.—Gas Analysis.
- " XI.—On the Capital necessary for starting a Carbon Works and the Profits in Carbon Manufacturing.
- " XII.—The Manufacture of Electrodes on a Small Scale.
- " XIII.—Building a Carbon Factory.
- " XIV.—Soot or Lamp Black.
- " XV.—Soot Factories.

Fully Illustrated. Price **7s. 6d.**, *post free,*

THE INCANDESCENT LAMP & ITS MANUFACTURE.
By GILBERT S. RAM.

190 pages, 116 illustrations. Price **3s. 6d.**, *post free.*

THE STEAM-ENGINE INDICATOR & INDICATOR DIAGRAMS.
Edited by W. W. BEAUMONT, M.I.C.E., M.I.M.E., &c.

Strongly bound in Cloth, **2s. 6d.** *net.*

TEMPERATURE COEFFICIENTS OF "CONDUCTIVITY" COPPER.

The compilers of "TEMPERATURE COEFFICIENTS OF 'CONDUCTIVITY' COPPER" (Messrs. CLARK, FORDE and TAYLOR, the well-known consulting engineers) have recently made fresh investigations of the temperature coefficient over the range of temperature to which Cables are usually exposed and the results of their investigations are now published.

- **Table I.** gives the resist. at different temperatures when the Res. at 32°F.=1.
- **Table II.** ditto ditto ditto at 60°F.=1.
- **Table III.** ditto ditto ditto at 75°F.=1; and
- **Table IV.** gives the log. reciprocals of co-efficients in Table III.

Also on strong cardboard, price **6d.** *net.*
SHEET TABLE showing
LOG. RECIPROCALS of COEFFICIENTS for COPPER RESISTANCES
At different temperatures from 32°F. to 84·9°F.

NEW VOLUMES IN PREPARATION.

NEARLY READY.

"THE ELECTRICIAN" WIREMAN'S POCKET BOOK.
Edited by F. CHARLES RAPHAEL.

A valuable collection of Tables and Instructions indispensable to those engaged in internal or external wiring work, whether as wiremen, superintendents or contractors. The volume will be of a size convenient for the pocket, and will contain a number of practical conductor and fall of potential tables, illustrated directions for joining wires and cables, diagrams of connections, wiring rules, tables of casings, tubing, &c., fuse tables, descriptions of all the various systems of wiring, Instructions for testing, &c. The greater part of the contents has been written especially for the book, and all has been thoroughly revised and brought up to date. The section on **Underground Work** will form a special feature, and will contain a mass of information hitherto unpublished. No pains are being spared to make the work as complete and useful as possible, and to give it an essentially **Practical** character.

THE ELECTRIC ARC. Fully illustrated. By Mrs. AYRTON. *In the Press.*
SECONDARY BATTERIES: THEIR CONSTRUCTION AND USE. By E. J. WADE.
PRIMARY BATTERIES: THEIR CONSTRUCTION AND USE. By W. R. COOPER. *In the Press.*
COMMERCIAL TELEPHONY. By DANE SINCLAIR and F. C. RAPHAEL. Fully Illustrated.

"The Electrician" Printing and Publishing Co., Ltd.

"THE ELECTRICIAN" SERIES—*continued.*

LARGE-SHEET TABLES,

Giving full particulars of the Electricity Supply Stations throughout the United Kingdom up to January, 1900, can be obtained mounted on rollers for hanging or folded into cloth cases. Tables (at foot of large sheet Table) give analyses of Electricity Supply undertaking Costs, and a neat coloured Map shows the positions of Provincial Supply Stations and Systems of Supply.

The Electric Railways and Tramways of the United Kingdom.
Similarly mounted for hanging or folded into cloth cases. PRICES ON APPLICATION.

Fully illustrated. Price 1s. 6d., *post free* 1s. 9d.

THE MANUFACTURE OF ELECTRIC LIGHT CARBONS.
A Practical Guide to the Establishment of a Carbon Manufactory.

The International Telegraph Convention and Service Regulations.
BUDAPEST REVISION (1896).

In the Official French Text with English Translation in parallel columns, by C. E. J. Twisaday (of the India Office, London), and Geo. R. Neilson (of the Eastern Telegraph Company, London). Cloth (foolscap folio), 5s. nett; wide margin (demy folio), 7s. 6d. nett. With Supplements, 1s. extra.

NOW READY.—Price £1. 5s., *in Sepia or Black; in very Massive Frame,* £2. 2s.

A Handsome Plate Reproduction of ROBERT DUDLEY's famous Painting of

THE "GREAT EASTERN"
(By permission of the Executors of the late Sir JAMES ANDERSON).

The subject measures 24in. by 17in., and is India mounted on fine etching boards, the mount measuring 27in. by 21in. The entire plate measures 30in. by 27in.

STEEL-PLATE PORTRAITS
OF

WILLOUGHBY SMITH (*out of print*),
MICHAEL FARADAY (1s. *extra*),
SIR JOHN PENDER, G.C.M.J.,
*LORD KELVIN, F.R.S.,
O. H. B. PATEY, C.B.
(Late Secretary of Telegraphs, G.P.O.),
DR. OLIVER J. LODGE, F.R.S.,
*SIR WILLIAM CROOKES, F.R.S.,
HERMANN VON HELMHOLTZ,
*PROF. W. E. AYRTON,
LORD RAYLEIGH, F.R.S.,
CYRUS W. FIELD,

*SIR W. H. PREECE, K.C.B., F.R.S.,
WERNER VON SIEMENS,
SIR JAMES ANDERSON, J.P.,
*ALEXANDER SIEMENS,
HEINRICH HERTZ,
*R. E. B. CROMPTON,
WILLIAM STURGEON,
PROF. J. J. THOMSON, F.R.S.,
*DR. JOHN HOPKINSON, M.A., F.R.S.,
*SIR HENRY C. MANCE, C.I.E., and
*JOSEPH WILSON SWAN

* Past-Presidents of the Institution of Electrical Engineers.

HAVE BEEN ISSUED AS SUPPLEMENTS TO

"𝕮𝖍𝖊 𝕰𝖑𝖊𝖈𝖙𝖗𝖎𝖈𝖎𝖆𝖓."

COPIES OF THESE ADMIRABLY-EXECUTED STEEL ENGRAVINGS CAN BE SUPPLIED, Price ONE SHILLING EACH; post free on Roller, 1s. 2d. India mounts, 1s. extra.

Or framed in neat Black Pillar or Brown Ornamental Frames, price
FOUR SHILLINGS AND SIXPENCE; carriage paid (U.K.), FIVE SHILLINGS.
if with mount, 1s. *extra.*

SECOND EDITION NOW READY.—Large Quarto, 370 pages, 140 full-page Illustrations, 65 full-page Tables, 30s.

ARMATURE WINDINGS OF ELECTRIC MACHINES.
By H. F. PARSHALL and H. M. HOBART.

The subjects dealt with in this work include the windings of continuous-current machines and armatures of single-phase, two-phase and three-phase machines, of induction motors for various phases, and of alternating current commutating machines. A chapter is given on electromotive force, explaining the properties of the different windings in this respect, and includes a discussion on the Sayers' armature windings.

The book will be found of great value to specialists, and is written in such a manner as to prove of considerable assistance to students and those desirous of thoroughly grasping the principles of dynamo construction. The exhaustive list of Tables dealing with continuous-current armature windings will prove particularly valuable at this time, as they show at a glance the arrangement possible for giving a number of conductors and poles.

"The Electrician" Printing and Publishing Co., Ltd.,

Demy 8vo., Cloth, fully illustrated, 10s. 6d., *post free* 11s.

THE APPLICATION OF ELECTRICITY TO RAILWAY WORKING.

By W. E. LANGDON, M.I.E.E.,
Superintendent and Engineer of the Electrical Department of the Midland Railway Company.

This work contains sections on the construction and surveying of Railway Telegraphs, Construction of Telegraph Instruments and Batteries, Block Signalling, Single-Line Working, Automatic Block Signalling, Interlocking, Miscellaneous Appliances in connection with Block Signalling, Electric Light and Power, Train Lighting, Intercommunication in Trains. The work also contains chapters relating to the engineering and traffic branches of railways, as well as specifications for telegraph poles, galvanised iron wire, copper wire, stranded laying wire, &c.

1899-1900 Edition (4th Year). 12s. 6d. *net.*

MANUAL OF ELECTRICAL UNDERTAKINGS.

By EMILE GARCKE, M.I.E.E., F.S.S.

The MANUAL contains financial particulars, &c., of all Electrical Companies and Undertakings. In the majority of cases the complete accounts of the various undertakings are shown, including the last Balance Sheet and Profit and Loss Account and similar data, together with much information showing the progress of the works. Particulars are given of all Municipal Electric Supply Stations

ELECTRICITY AND MAGNETISM.

A TREATISE ON MAGNETISM AND ELECTRICITY. By Prof A Gray. In 2 vols. Vol. I., 14s. net.

AN ELEMENTARY TREATISE ON FOURIER'S SERIES, and Spherical, Cylindrical and Ellipsoidal Harmonics, with Applications to Problems on Mathematical Physics. By Prof. Byerly, Harvard University. 12s. 6d.

MODERN VIEWS OF ELECTRICITY. By Oliver J. Lodge, F.R.S., Professor of Physics in University College, Liverpool. Illustrated. 6s. 6d.

THE ELECTRO-MAGNET AND ELECTROMAGNETIC MECHANISM. By Silvanus P. Thompson, D.Sc., F.R.S. 2nd Edition, 450 pages, 213 illustrations. 15s.

ELECTRICITY: ITS THEORY, SOURCES AND APPLICATIONS. By John T. Sprague. Third Edition. Revised and enlarged. 15s.

ELECTRICAL PAPERS. For Advanced Students in Electricity. By Oliver Heaviside. 2 vols. 12s. net. (Published 31s. 6d.)

A COURSE OF LECTURES ON ELECTRICITY, DELIVERED BEFORE THE SOCIETY OF ARTS. By George Forbes, M.A., F.R.S. (L. & E.) With 17 illustrations, crown 8vo. 5s.

MAGNETS AND ELECTRIC CURRENTS: An Elementary Treatise for Electrical Artisans and Science Teachers. By Dr. J. A. Fleming, M.A., F.R.S., &c. Crown 8vo., Cloth 7s. 6d.

LECTURES IN ELECTRICITY AT THE ROYAL INSTITUTION, 1875-76 By John Tyndall. 2s. 6d.

NOTES OF A COURSE OF SEVEN LECTURES ON ELECTRICAL PHENOMENA. By John Tyndall. 1s. 6d.

ELECTRIC WAVES: Being Researches on the Propagation of Electric Action with Finite Velocity through Space. By Dr. Heinrich Hertz. Translated by D. E. Jones. 10s. 6d. net.

ELECTRICAL MEASUREMENTS. By Prof. H. S. Carhart and G. W. Patterson, jun. 8s. 6d.

ELECTRIC MOVEMENTS IN AIR AND WATER. By Lord Armstrong, C.B., F.R.S. New Edition. 30s. net.

LECTURES ON SOME RECENT ADVANCES IN PHYSICAL SCIENCE. By Prof. P. G. Tait. Third Edition. 9s.

ELEMENTARY LESSONS IN ELECTRICITY AND MAGNETISM. By Prof. Silvanus P. Thompson, D.Sc., F.R.S. New Edition. Fcap. 8vo., 4s. 6d.

PRACTICAL ELECTRICITY: A Laboratory and Lecture Course for First Year Students of Electrical Engineering. By Prof. W. E. Ayrton, F.R.S. New Edition. Vol. I., Current, Pressure, Resistance, Energy, Power and Cells. 247 Illustrations. 9s.

A TREATISE ON ELECTRICITY AND MAGNETISM. By J. Clerk Maxwell, M.A., F.R.S. Third Edition. 2 vols., demy 8vo, cloth, £1. 12s.

CENTENARY OF THE ELECTRIC CURRENT, 1799-1899—*see page 13.*

ELEMENTARY TREATISE ON ELECTRICITY AND MAGNETISM. Founded on Joubert's "Traité Élémentaire d'Electricité. By Prof. G. C. Foster, F.R.S., and E. Atkinson, Ph.D. Crown 8vo, 7s. 6d.

MAGNETISM AND ELECTRICITY, AN ELEMENTARY MANUAL OF. With Examination Questions and many illustrations. By Prof. A. Jamieson. Fourth Edition. Cr. 8vo, 3s. 6d.

1, 2 and 3, Salisbury Court, Fleet Street, London, E.C.

AN ELEMENTARY TREATISE ON ELECTRICITY. By J. Clerk Maxwell.
M.A., F.R.S. Edited by William Garnett, M.A. Demy 8vo, cloth. 7s. 6d.
RECENT RESEARCHES IN ELECTRICITY AND **MAGNETISM.** By
Prof. J. J. Thomson, M.A., F.R.S. 18s. 6d.
THE THEORY OF ELECTRICITY AND MAGNETISM. By Prof. A. G. Webster. 14s.
COIL AND CURRENT, OR THE TRIUMPHS OF ELECTRICITY. By
Henry Frith and W. Stepney Rawson. 3s. 6d.
MAGNETISM AND ELECTRICITY. By W. J. Harrison and C. A. White. 2s.
MAGNETISM AND ELECTRICITY. By A. W. Poyser, M.A. Cloth, 2s. 6d.
MAGNETISM AND ELECTRICITY. By H. C. Tarn, F.S.Sc.
MAGNETISM AND ELECTRICITY. By J. Paley Yorke. 3s. 6d.
MAGNETISM AND ELECTRICITY. By R. Wallace Stewart, D.Sc. 159 Illustrations. Second Edition. 5s. 6d.
MAGNETISM AND ELECTRICITY. By Edward Aveling, D.Sc. 6s.
MAGNETISM AND ELECTRICITY. By John Angell. 2s.
MAGNETISM AND ELECTRICITY. By F. Guthrie, B.A., Ph.D. 3s. 6d.
MAGNETISM AND ELECTRICITY FOR BEGINNERS. By H. E. Hadley, B.Sc. 2s. 6d.
MANUAL OF ELECTRICAL SCIENCE. By George J. Burch, B.A. 3s.
ADVANCED ELECTRICITY AND MAGNETISM. By A. W. Poyser. 4s. 6d.
FIRST BOOK OF ELECTRICITY AND MAGNETISM. By W. Perren Maycock. 107 Illustrations. Second Edition. 2s. 6d.
ELECTROMAGNETIC THEORY—see page 4.
A.B.C. OF ELECTRICITY. By W. H. Meadowcroft. Fourth Edition. 2s.
ELECTRICITY AND MAGNETISM. Translated from the French of Prof.
E. Gerard, under the direction of Dr. Louis Duncan, with additions by Dr. Louis Duncan, C. P. Steinmetz, A. E. Kennelly and Dr. Cary T. Hutchinson. Cloth, 10s. 6d.
ELEMENTS OF THE MATHEMATICAL THEORY OF ELECTRICITY AND
MAGNETISM. By J. J. Thomson, F.R.S. 10s.
ELEMENTS OF STATIC ELECTRICITY. By P. Atkinson. Second Edition. 7s.
ELECTRICITY TREATED EXPERIMENTALLY, for the use of schools and
students. By Linnaeus Cumming, M.A. Fifth Edition. 4s. 6d.
**THE ALTERNATE-CURRENT TRANSFORMER IN THEORY AND
PRACTICE** see pages 6 and 7.
ELECTRICITY AND MAGNETISM. By S. R. Bottone. 3s. 6d.
BOOK E. ARITHMETICAL PHYSICS. Part IIA.—MAGNETISM AND ELECTRICITY, ELEMENTARY AND ADVANCED. With Supplement on Lines of Force. By C. J. Woodward, B.Sc. 2s.
BOOK F. ARITHMETICAL PHYSICS. Part IIB.—MAGNETISM AND ELECTRICITY, DEGREE AND HONOURS STAGES. By C. J. Woodward, B.Sc. New Edition. 3s. 6d.
MAGNETIC INDUCTION IN IRON AND OTHER METALS—see page 4.
ELECTRICITY FOR PUBLIC SCHOOLS AND COLLEGES. By W. Larden, M.A. 6s.
ELECTRICITY. By Dr. Ferguson; revised and extended by Prof. J. Blyth. 3s. 6d.
ELECTRICITY AND MAGNETISM. By Fleeming Jenkin. 3s. 6d.
ELECTRICITY AND MAGNETISM. By Prof. F. E. Nipher. Second Edition. 14s.
ELECTRICITY AND MAGNETISM. By Prof. Balfour Stewart, F.R.S., and
W. W. Haldane Gee. Crown 8vo, 7s. 6d.; School Course, 2s. 6d.
ELECTRICITY AND MAGNETISM. A Popular Treatise. By Amédée
Guillemin. Translated by Silvanus P. Thompson. 31s. 6d.
ELEMENTARY TREATISE ON PHYSICS, Experimental and Applied. Translated from Ganot's "Elements de Physique," by E. Atkinson, Ph.D. Twelfth Edition. 15s.
ELEMENTARY PHYSICS. By John Henderson. 2s. 6d.
PRACTICAL ELECTRICITY AND MAGNETISM. By John Henderson. 6s. 6d.
PHYSICS. Advanced Course. By G. F. Barker. 21s.
THE INTELLECTUAL RISE IN ELECTRICITY. By Park Benjamin, Ph.D. 21s.
TEXT-BOOK OF THE PRINCIPLES OF PHYSICS. By Alfred Daniell.
Second Edition. 21s.
PHYSICAL UNITS. By Magnus Maclean, M.A., D.S.C., F.R.S.E. 2s. 6d.

ABSOLUTE MEASUREMENTS IN ELECTRICITY AND MAGNETISM. By Prof. Andrew Gray. Second Edition. 5s. 6d.
ALTERNATING CURRENTS OF ELECTRICITY. By T. H. Blakesley. Third Edition. 5s.
ALTERNATING CURRENTS OF ELECTRICITY: Their Generation, Measurement, Distribution, and Application. By Gisbert Kapp. 4s. 6d.
ALTERNATING CURRENTS. An Analytical and Graphical Treatment for Students and Engineers. By Dr. F. Bedell and Dr. A. C. Crehore. Third Edition. 11s.
ALTERNATING CURRENTS OF ELECTRICITY AND THE THEORY OF TRANSFORMERS. By Alfred Still. 5s.
ARITHMETIC OF MAGNETISM AND ELECTRICITY. By R. Gunn. 2s. 6d.
PHYSICAL ARITHMETIC. By Alexander Macfarlane. 7s. 6d.
THE ALTERNATING CURRENT CIRCUIT. By W. Perren Maycock. 2s. 6d.
QUESTIONS AND EXAMPLES IN EXPERIMENTAL PHYSICS, SOUND, LIGHT, HEAT, ELECTRICITY, AND MAGNETISM. By B. Loewy. 2s.
LESSONS IN ELEMENTARY PHYSICS. By Prof. B. Stewart. Fifth Edition, 4s. 6d.
ELECTRICITY AND MAGNETISM FOR BEGINNERS. By F. W. Sanderson, M.A. 2s. 6d.
ELECTRICITY AND MAGNETISM. By L. Cumming, M.A. 2s. 6d.
THE THEORY AND CALCULATION OF ALTERNATING CURRENT PHENOMENA. By C. P. Steinmetz. Second Edition. 11s.
HANDBOOK OF ELECTRICITY, MAGNETISM, AND ACOUSTICS. By Dr. Lardner. Edited by Geo. Carey Foster, B.A. 5s.
THEORY OF ELECTRICITY AND MAGNETISM. By C. E. Curry, Ph.D., with Preface by Prof. L. Boltzmann. 8vo., 400 pages, 8s. 6d. nett.
INTRODUCTION TO THE THEORY OF ELECTRICITY. By Linnæus Cumming, M.A. Fourth Edition. 8s. 6d.
PAPERS ON ELECTRO-STATICS AND MAGNETISM. By Lord Kelvin. Second Edition. 18s.
PRACTICAL NOTES FOR ELECTRICAL STUDENTS—*see page 13.*
PRACTICAL MEASUREMENT OF ELECTRICAL RESISTANCE. By W. A. Price, M.A. 14s.
POLYPHASED ALTERNATING CURRENTS. By E. Hospitalier. 3s. 6d.
MATHEMATICAL THEORY OF ELECTRICITY AND MAGNETISM. Vol. I., Electrostatics; Vol. II., Magnetism and Electrodynamics. By H. W. Watson, D.Sc., and S. H. Burbury, M.A. 8vo, 10s. 6d. each.
MATHEMATICAL THEORY OF ELECTRICITY AND MAGNETISM, AN INTRODUCTION TO. By W. T. A. Emtage, M.A. Second Edition. Crown 8vo, 7s. 6d.
ELECTRICITY: A Sketch for General Readers. By E. M. Caillard. 7s. 6d.
ELEMENTS OF DYNAMIC ELECTRICITY AND MAGNETISM. By Philip Atkinson, A.M., Ph.D. 10s. 6d.
ELEMENTS OF PHYSICS. By Prof. H. S. Carhart. 5s.
PHYSICS, A TEXT-BOOK OF. By W. Watson. 10s. 6d.
ELECTRICAL ENGINEERING FORMULÆ—*see page 5.*
THE ARITHMETIC OF ELECTRICAL MEASUREMENTS. By W. R. P. Hobbs, Head Master of the Torpedo School, H.M.S. "Vernon." New Edition. 1s.
THEORY AND PRACTICE OF ABSOLUTE MEASUREMENTS IN ELECTRICITY AND MAGNETISM. By Andrew Gray, M.A., F.R.S.E., Professor of Physics in the University College of North Wales. In 2 vols., crown 8vo. Vol. I., 12s. 6d. Vol. II., in 2 parts, 25s.
PROBLEMS AND SOLUTIONS IN ELEMENTARY ELECTRICITY AND MAGNETISM. By W. Slingo and A. Brooker. 2s.
THE CAPILLARY ELECTROMETER IN THEORY AND PRACTICE. By George J. Burch, M.A. Paper covers, 2s., net, strong cloth, 2s. 6d., net.
EXERCISES IN ELECTRICAL AND MAGNETIC MEASUREMENTS, with Answers. By R. E. Day. 3s. 6d.
INTRODUCTION TO PHYSICAL MEASUREMENTS. By Dr. F. Kohlrausch. Translated by T. H. Waller and H. R. Procter. Third English Edition. 15s.
ELECTRICAL MEASUREMENT AND THE GALVANOMETER, ITS CONSTRUCTION AND USES. By T. D. Lockwood. Second Edition. 6s.
ELECTRIC CURRENT: How Produced and How Used. By R. Mullineux Walmsley, D.Sc. 10s. 6d.

"**THE ELECTRICIAN" PRIMERS**—*see page 8.*
ELECTRICITY, ELECTROMETER, MAGNETISM AND ELECTROLYSIS.
By G. Chrystal and W. N. Shaw. 5s. nett.
INTERMEDIATE COURSE OF PRACTICAL PHYSICS. By Prof. A.
Schuster and Charles H. Lees, D.Sc. 5s.
ELECTRICITY IN THEORY AND PRACTICE. By Lieut. B. A. Fiske. 10s. 6d.
ELECTRICITY FOR ENGINEERS. By Ch. Desmond. Revised Edition. 10s. 6d.
POTENTIAL: Its Application to the Explanation of Electrical Phenomena Popularly Treated. By Dr. Turnlirz. Translated by D. Robertson, M.A. 3s. 6d.
ELECTRICAL INFLUENCE MACHINES: containing a full account of their Historical Development, their Modern Forms, and their Practical Construction. By John Gray, B.Sc 89 Illustrations. 4s. 6d.
WHAT IS ELECTRICITY? By John Trowbridge, S.D. Crown 8vo, 5s
UNITÉS ÉLECTRIQUES ABSOLUES. By M. G. Lippmann. 8s. 6d.
ELECTRICITY IN THE SERVICE OF MAN. A Popular and Practical Treatise on the Applications of Electricity in Modern Life. New Edition. Revised by R. Mullineux Walmsley, D.Sc. (Lond.), F.R.S.E. Medium 8vo., with upwards of 950 illustrations. 10s. 6d. New and Cheap Edition, 7s. 6d.
DOMESTIC ELECTRICITY FOR AMATEURS. Translated from the French of E. Hospitaller, with additions By C. J. Wharton, M.I.E.E. Numerous illustrations, demy 8vo, cloth, 6s.
PRACTICAL ELECTRICAL MEASUREMENTS By Ellis H. Crapper. 2s. 6d.
PRACTICAL ELECTRICS: A Universal Handy Book on Every-Day Electric Matters. Third Edition. 3s. 6d.
THE WORK OF HERTZ.—*see page 13.*
PRACTICAL MAGNETISM AND ELECTRICITY. By J. R. Ashworth. 2nd Edition. 2s. 6d.
ELECTRICITY: A Hundred Years Ago and To-Day. By Prof. E. J. Houston. 4s. 6d.
ELECTRICITY IN MINING. By Silvanus P. Thompson, D.Sc., F.R.S. 2s.
A CENTURY OF ELECTRICITY. By Prof. T. C. Mendenhall. 4s. 6d.
LABORATORY MANUAL OF PHYSICS AND APPLIED ELECTRICITY.
By Prof. Edward L. Nichols. 12s. 6d.
STANDARD METHODS IN PHYSICS & ELECTRICITY CRITICISED, AND A TEST FOR ELECTRIC METERS PROPOSED. By H. A. Naber. Demy 8vo., cloth gilt, 6s., post free.
LEÇONS SUR L'ÉLECTRICITE ET LE MAGNETISME. By MM. E. Mascart and J. Joubert. Two volumes. Second Edition, 30s. Each volume, 15s.
TRAITÉ ÉLÉMENTAIRE D'ÉLECTRICITÉ. By J. Joubert. 6s.
THÉORIE DE L'ÉLECTRICITE. By Prof. A. Vaschy. 16s.
LEÇONS SUR L'ÉLECTRICITE. By Eric Gerard. 2 vols., fully illustrated. 6th Edition. 19s. In 2 Vols., 10s. each.
TRAITÉ PRATIQUE DE L'ELECTRICITÉ. By Felix Lucas. 13s 6d.
TRAITÉ D'ELECTRICITÉ: Theorie et Applications Générales. By F. Rodary. 16s. 6d.
THÉORIE DE POTENTIEL NEWTONIEN. By H. Poincaré. 12s.

ELECTRIC LIGHTING & TRANSMISSION OF POWER.
THE ALTERNATE-CURRENT TRANSFORMER IN THEORY AND PRACTICE—*see page 6.*
ALTERNATING ELECTRIC CURRENTS. By Prof. E. J. Houston, Ph.D., and A. E. Kennelly, D.Sc. 4s. 6d.
DYNAMO-ELECTRICITY: Its Generation, Application, Transmission, Storage and Measurement. By G. B. Prescott. 545 Illustrations. £1. 1s.
DYNAMO MACHINERY AND ALLIED SUBJECTS (ORIGINAL PAPERS ON). By Dr. John Hopkinson. 5s.
ELECTRO DYNAMICS: The Direct Current Motor. By Prof. C. A. Carus-Wilson. 7s. 6d.
"**THE ELECTRICIAN" PRIMERS**—*see page 8.*
STANDARD POLYPHASE APPARATUS AND SYSTEMS. By Maurice E Oudin, M.S. 13s.
SHOP AND ROAD TESTING OF DYNAMOS AND MOTORS. By Eugene C Parham and John C. Shedd. 9s.

ARMATURE WINDINGS OF ELECTRIC MACHINES. New Ed. By H. F.
Parshall and H. M. Hobart. Large 4to. 140 illustrations and 65 full-page tables. 30s. post free.—*See page 15*

DRUM ARMATURES AND COMMUTATORS—*see page 13.*

ELECTRIC ARC LIGHTING. By Prof. E. J. Houston, Ph.D., and A. E.
Kennelly, D.Sc. Cloth, 4s. 6d.

ELECTRO MAGNETISM. By Prof. E. J. Houston, and A. E. Kennelly. 4s. 6d.

CARBON MANUFACTURE FOR ALL ELECTRICAL PURPOSES. By
Francis Jehl. *See page 14.*

CONTINUOUS-CURRENT DYNAMOS AND MOTORS. An Elementary
Treatise for Students. By Frank P. Cox. 9s.

ELECTRIC HEATING. By Prof. E. J. Houston and A. E. Kennelly. Cloth, 4s. 6d.

THE PRACTICAL APPLICATION OF DYNAMO ELECTRICAL MACHI-
NERY. By C. K. MacFadden and Wm. D. Ray. 4s. 6d.

PRACTICAL NOTES FOR ELECTRICAL STUDENTS—*see page 13.*

HOW TO BUILD DYNAMO-ELECTRIC MACHINERY. By Ed. Trevert. 10s. 6d.

THE INCANDESCENT LAMP AND ITS MANUFACTURE—*see page 14.*

ELECTRIC INCANDESCENT LIGHTING. By Prof. E. J. Houston, Ph.D.,
and A. E. Kennelly, D.Sc. Cloth, 4s. 6d.

THE ELECTRIC MOTOR. By Prof. E. J. Houston and A. E. Kennelly. 4s. 6d.

DIGEST OF THE LAW OF ELECTRIC LIGHTING, &c.—*see page 8.*

DISSECTED MODEL OF A DIRECT-CURRENT DYNAMO, with a Description of the Various Parts and an Explanation of their Mode of Action, and the Method of Winding a Drum Armature. By Arnold Philip. 4s. 6d.

DYNAMO CONSTRUCTION: A Practical Handbook for Engineer Constructors
and Electricians in Charge. By John W. Urquhart. Second Edition. 7s. 6d.

DYNAMO-ELECTRIC MACHINERY: A Text-Book for Students of Electro-
Technology. By Silvanus P. Thompson, B.A., D.Sc., M.I.E.E., F.R.S. Fifth Edition, revised and enlarged. Cloth, 864 pages, 29 folding plates, 498 illustrations in text. 24s. post free.

DYNAMO AND MOTOR BUILDING FOR AMATEURS. By C. D.
Parkhurst. Cloth, 5s.

DYNAMO, THE: Its Theory, Design and Manufacture. By C. C. Hawkins,
A.M.Inst.C.E., and F. Wall's. Second Edition. 10s. 6d.

PRACTICAL MANAGEMENT OF DYNAMOS AND MOTORS. By F. B.
Crocker and S. S. Wheeler. Second Edition. 4s. 6d.

ELECTRICAL ENGINEERING FORMULÆ—*see page 5.*

ELECTRO-DYNAMIC MACHINERY. By Prof. E. J. Houston, Ph.D., and
A. E. Kennelly, D.Sc. 232 Illustrations. Cloth, 10s. 6d.

TRANSFORMERS FOR SINGLE AND MULTIPHASE CURRENTS. By
Gisbert Kapp. Translated from the German by the Author. 6s.

TEXT BOOK ON ELECTRO-MAGNETISM AND THE CONSTRUCTION
OF DYNAMOS. By Prof. D. C. Jackson. Vol. I., Electro-magnetism and the Construction of Continuous-current Dynamos. 9s. net; Vol. II., Alternating Currents and Alternating Current Machinery. 14s. net

THE MANUFACTURE OF ELECTRIC LIGHT CARBONS—*see page 14.*

ELECTRICAL POWER TRANSMISSION. By Louis Bell, Ph.D. 11s.

PRACTICAL DIRECTIONS FOR WINDING MAGNETS FOR DYNAMOS.
By Carl Hering. 3s. 6d.

THE STEAM ENGINE INDICATOR, AND INDICATOR DIAGRAMS—
see page 15.

ON THE DEVELOPMENT AND TRANSMISSION OF POWER FROM
CENTRAL STATIONS. By Prof. W. C. Unwin. 10s. nett.

ELEMENTS OF CONSTRUCTION FOR ELECTRO-MAGNETS. By Count du
Moncel. Translated by C. J. Wharton. 4s. 6d.

APPLIED MAGNETISM. By J. A. Kingdon, B.A. 7s. 6d.

ON THE CALCULATION OF NETWORKS FOR DISTRIBUTION OF
POWER BY CONTINUOUS AND ALTERNATING CURRENTS. By H. Cohen. In German. 2s. 6d.

ELECTRIC LIGHTING AND POWER DISTRIBUTION: An Elementary
Manual for Students. By W. Perren Maycock. In two vols. Vol. I. (Fourth Edition) 6s.

POLYPHASE ELECTRIC CURRENTS AND ALTERNATE CURRENT
MOTORS. By Prof. Silvanus P. Thompson, D.Sc. New Edition. Demy 8vo. Cloth 21s.

CONTINENTAL ELECTRIC LIGHT CENTRAL STATIONS. By Killingworth
Hedges. 15s.

CENTRAL STATION ELECTRICITY SUPPLY. By Albert Gay and C. H. Yeaman. 10s. 6d.
PRACTICAL CALCULATIONS OF DYNAMO-ELECTRIC MACHINES. By Alfred E. Wiener. 11s.
THEORETICAL ELEMENTS OF ELECTRO-DYNAMIC MACHINERY. Vol. I. By A. Kennelly, F.R.A.S. 4s. 6d.
THE DYNAMO-TENDER'S HANDBOOK. By F. B. Badt. 4s. 6d.
ELECTRIC LIGHTING. By Francis B. Crocker. Vol. I.: Generating Plant. 12s. 6d.
NOTES ON DESIGN OF SMALL DYNAMOS. By George Halliday. 2s. 6d.
ELECTRIC LIGHT INSTALLATIONS. By Sir David Salomons, Bart. In Three Volumes. Vol. I., 5s.: The Management of Accumulators. Vol. II., 7s. 6d.: Apparatus, Engines, Dynamos and Motors, Instruments, Governors, Switches and Switch Boards, Fuses, Cut-Outs, Connectors and Minor Apparatus, Arc Lamps, Practical Applications. Vol. III., 5s.: Application
POWER DISTRIBUTION FOR ELECTRIC RAILROADS. By Dr. L. Bell. 11s.
ELECTRICAL BOATS AND NAVIGATION. By T. Commerford Martin and Joseph Sachs. Cloth 8vo, 232 pages, 148 illustrations. 10s. 6d.; post free, 11s
A HANDBOOK OF ELECTRICAL TESTING. By H. R. Kempe, M.I.E.E. Fifth Edition, revised and enlarged. 18s.
THE MAGNETIC CIRCUIT IN THEORY AND PRACTICE. By Dr. H. du Bois. Translated by Dr. E. Atkinson. 12s. nett.
ELECTRIC LIGHT: Its Production and Use. By John W. Urquhart. Sixth Edition, carefully revised, with large additions. 7s. 6d.
ELECTRICAL ENGINEERING: For Electric Light Artizans and Students. By W. Slingo and A. Brooker. Revised and enlarged edition. 12s.
LOCALISATION OF FAULTS IN ELECTRIC LIGHT MAINS—*see page 10.*
DISSECTED MODEL OF A DIRECT-CURRENT DYNAMO, with a Description of the Various Parts and an Explanation of their Mode of Action and the Method of Winding a Drum Armature. By Arnold Philip, B.A., B.Sc. 4s. 6d.
THE ELECTRICAL TRANSMISSION OF ENERGY. By A. V. Abbott. Second Edition, 608 pages, with eight folding plates and numerous illustrations. 21s. nett.
ELECTRIC SHIP LIGHTING: For the use of Ship Owners and Builders Engineers, &c. By John W. Urquhart. 7s. 6d.
THE PHŒNIX FIRE OFFICE RULES FOR ELECTRIC LIGHT INSTALLATIONS AND ELECTRICAL POWER INSTALLATIONS. By Musgrave Heaphy, C.E. Twenty-seventh Edition, 8vo, sewed. 6d.
THE MANAGEMENT OF DYNAMOS. By G. Lummis-Paterson. New Edition. 4s. 6d.
ELECTRIC LIGHTING FOR MARINE ENGINEERS. By S. F. Walker. 5s.
COLLIERY LIGHTING BY ELECTRICITY. By S. F. Walker. 2s. 6d.
ELECTRIC LAMPS AND ELECTRIC LIGHTING—*see page 5.*
HOW TO MAKE A DYNAMO. By Alfred Crofts. 5th Edition. Cloth, 2s.
THE DYNAMO: HOW MADE AND HOW USED. By S. R. Bottone. With 39 illustrations Tenth Edition. 2s. 6d.
TREATISE ON INDUSTRIAL PHOTOMETRY, with Special Application to Electric Lighting. By Dr. A. Palaz. Translated by G. W and M. R. Patterson. 12s. 6d.
MAY'S POPULAR INSTRUCTOR FOR THE MANAGEMENT OF ELECTRIC LIGHTING PLANT. An indispensable Handbook for persons in charge of Electric Lighting plants, more particularly those with slight technical training. Pocket size, price 2s. 6d.; post free, 2s. 8d
ELECTRIC LIGHT ARITHMETIC. By R. E. Day. 2s.
THE POTENTIOMETER AND ITS ADJUNCTS.—*See page 10.*
THE GALVANOMETER AND ITS USES. A Manual for Electricians and Students. By C. H. Haskins. Second edition. Illustrated. 18mo., 7s.
TRANSFORMERS: Their Theory, Construction, and Application Simplified. By Caryl D. Haskyns. 4s. 6d.
MAY'S BELTING TABLE. Showing the relations between—(1) The number of revolutions and diameter of pulleys and velocity of belts; (2) The horse-power, velocity, and square section of belts; (3) The thickness and width of belts; (4) The square section of belts at different strains per square inch. For office use, printed on cardboard, with metal edges and suspender, price 2s.; post free, 2s. 2d.; for the pocket, mounted on linen, in strong case, 2s. 6d.; post free, 2s. 8d.
ALTERNATE CURRENTS IN PRACTICE. Translated from the French of Loppé and Bouquet by F. J. Moffett. 15s.

1, 2 and 3, Salisbury Court, Fleet Street, London, E.C.

ELECTRICITY IN TOWN AND COUNTRY HOUSES. By P. E. Scrutton. 2s. 6d.
INDUSTRIAL ELECTRICITY. Translated from the French of M. H. De Graffigny, and edited by A. G. Elliott. 2s. 6d.
EXAMPLES IN ELECTRICAL ENGINEERING. By S. Joyce, A.I.E.E. 5s.
ELECTRIC TRANSMISSION OF ENERGY, and its Transformation, Subdivision and Distribution. By Gisbert Kapp. Fourth Edition. 10s. 6d.
ELECTRIC LIGHT CABLES AND THE DISTRIBUTION OF ELECTRICITY. By Stuart A. Russell, Assoc.M.Inst.C.E. 107 illustrations 7s. 6d.
PRACTICAL ELECTRICAL NOTES AND DEFINITIONS, for the use of Engineering Students and Practical Men. By W. Perren Maycock. Second Edition. 3s.
ELECTRICAL ENGINEERS' AND STUDENTS' CHART AND HANDBOOK OF THE BRUSH ARC LIGHT SYSTEM. By H. C. Reagen, jun. Second Edition. 4s. 6d.
THE GALVANOMETER: A Series of Lectures. By Prof. E. L. Nichols. 4s. 6d.
ELECTRIC MOTIVE POWER—see page 9.
ELECTRO-MOTORS: How Made and How Used. By S. R. Bottone. 2nd ed. 3s.
THE ELECTRIC RAILWAY OF TO-DAY. By H. B. Prindle. 2s. 6d.
ELECTRIC TRANSMISSION HANDBOOK. By F. B. Badt. 4s. 6d.
AMERICAN ELECTRIC STREET RAILWAYS: Their Construction and Equipment. By Killingworth Hedges. 12s. 6d.
ELECTRIC RAILWAY MOTORS: Their Construction, Operation and Maintenance. By Nelson W. Perry. 4s. 6d.
ELECTRIC RAILWAYS AND TRAMWAYS: Their Construcion and Operation. Revised, enlarged and brought up to date from *Engineering*. By Philip Dawson, C.E. Demy 4to, handsomely bound in Half Morocco. 678 pages illustrated, and containing many Tables, Diagrams, and Working Drawings. 42s.
ELECTRIC TRACTION POCKET BOOK. By Philip Dawson, A M.I.C.E., M.I.E.E. Fcap. 8vo, 17s. 6d. net.
ELECTRICITY AS A MOTIVE POWER. By Count du Moncel and Frank Geraldy. Translated by C. J. Wharton. 7s. 6d.
ELECTRIC TRANSFORMATION OF POWER. By Philip Atkinson. 7s. 6d.
MOTIVE POWER AND GEARING—see page 9.
ELECTRIC RAILWAYS, RECENT PROGRESS IN. By Carl Hering. 5s.
ELECTRIC TRACTION. By Prof. Ernest Wilson. 5s.
ELECTRIC TRAMWAYS AND RAILWAYS POPULARLY EXPLAINED. By H. Scholey. 2s.
ELECTRIC RAILWAY ENGINEERING. By Edward Trevert. 8s.
ELECTRIC STREET RAILWAYS. By Prof. E. J. Houston, Ph.D., and A. E. Kennelly, D.Sc. Cloth, 4s. 6d.
THE WESTINGHOUSE AND BALDWIN SYSTEMS OF ELECTRIC LOCOMOTIVES. 6s.
PRIMARY BATTERIES. By H. S. Carhart, A.M. 67 illustrations. 6s.
ELECTRO-MOTORS: The Means and Apparatus employed in the Transmission of Electrical Energy and its Conversion into Motive Power. By John W. Urquhart. 7s. 6d.
THE STORAGE OF ELECTRICAL ENERGY, and Researches in the Effects created by Currents Combining Quantity with High Tension. By G. Planté. Translated from the French by Paul Bedford Elwell. With 89 illustrations, 8vo, cloth, 12s.
THE ELECTRIC MOTOR AND ITS APPLICATIONS. By T. C. Martin and J. Wetzler (with an Appendix on the Development of the Electric Motor since 1888, by Dr. Louis Bell). Third Edition. Quarto, 315 pages, 353 illustrations. 12s. 6d., post free 13s. 6d.
LA TRACTION ÉLECTRIQUE SUR VOIES FERRÉES. By Prof. A. Blondel and F. Paul-Dubois. Two vols. 42s.
DISTRIBUTION DE L'ÉNERGIE PAR COURANTS POLYPHASÉS. By J. Rodet. 7s.
LES DYNAMOS. By J. A. Montpellier. 12s. 6d.
TRAITÉ DES PILES ÉLECTRIQUES; Piles hydro-électriques—Accumulateurs—Piles thermo-électriques et pyro-électriques. By Donato Tommasi, Docteur-ès-Sciences. 8vo., 10s.
TRAITE COMPLET D'ELECTRO-TRACTION. By Ernest Gerard. 25s.
LES TRAMWAYS ÉLECTRIQUES. By H. Maréchal. 6s. 6d.

"The Electrician" Printing and Publishing Co., Ltd.,

DYNAMOMASCHINEN FÜR GLEICH UND WECHSELSTROM. By Gisbert Kapp. 12s. 6d.
ARC LAMPS AND HOW TO MAINTAIN THEM. By Smithson and Sharpe. 1s.
UNTERSUCHUNGEN UEBER DIE AUSBREITUNG DER ELEKTRISCHEN KRAFT. By the late Prof. H. Hertz. Price 6s. 6d.
ELEKTROMECHANISCHE KONSTRUKTIONEN. By Gisbert Kapp. 200 quarto pages, 25 tables, and 54 illustrations. 21s.
WIRKUNGSWEISE, BERECHNUNG UND KONSTRUKTION ELEKTRISCHER GLEICHSTROM-MASCHINEN. By J. Fischer-Hinnen. Fourth Edition, revised and enlarged, 14s.

WIRING AND ELECTRIC LIGHT FITTING.

"THE ELECTRICIAN" WIREMAN'S POCKET BOOK—see page 14.
THE INCANDESCENT LAMP AND ITS MANUFACTURE—see page 14.
ELECTRIC LIGHTING SPECIFICATIONS, for the Use of Engineers and Architects. By E. A. Merrill. 6s.
STANDARD WIRING FOR ELECTRIC LIGHT AND POWER. By H. C. Cushing, Jr. Strongly bound in cloth, 4s. 6d., post free.
ELECTRIC WIRING For the Use of Architects, Underwriters and the Owners of Buildings. By Russell Robb. 16s.
ELECTRIC LIGHT FITTING: A Handbook for Electrical Engineers. By John W. Urquhart. Fully Illustrated. 5s.
COMPREHENSIVE INTERNATIONAL WIRE TABLES. By W. S. Boult. Full particulars of 469 Conductors (4 gauges), Single Wires and Cables, in English, American, and Continental Units. Price 5s., post free. See page 4.
ELECTRIC LIGHT FOR COUNTRY MANSIONS. By John Henry Knight. 1s.
PRACTICAL ELECTRIC LIGHT FITTING. By F. C. Allsop. 5s.
WIRING SLIDE RULE (Trotter's Patent). By which can be found at once .—
1. Size of Cable; 2. Length of Cable; 3. Current Cable will carry; 4. Current Density; 5. Maximum Current; 6. Resistance in Ohms; 7. Sectional Area in Square Inches. Full printed instructions are supplied with each rule. For the pocket. Price 2s. 6d.; post free, 2s. 7d.
STANDARD TABLES FOR ELECTRIC WIREMEN. By Charles M. Davis. 5s.
MAY'S TABLE OF ELECTRIC CONDUCTORS. Showing the relations between—(1) The sectional area, diameter of conductors, loss of potential, strength of current, and length of conductors; (2) The economics of incandescent lamps, their candle-power, potential, and strength of current; (3) The sectional area, diameter of conductors, and strength of current per square inch. For office use, printed on cardboard, with metal edges and suspender, price 2s.; post free, 2s. 2d.; for the pocket, mounted on linen, in strong case, 2s. 6d.; post free, 2s. 8d.
UNIVERSAL WIRING COMPUTER. By Carl Hering. 5s.
INCANDESCENT WIRING HANDBOOK. With Tables. By F. B. Badt. 4s. 6d.
DOMESTIC ELECTRICITY FOR AMATEURS. By E. Hospitalier. Translated by C. J. Wharton. 6s.
HOW TO WIRE BUILDINGS. By Augustus Noll. Fourth Edition 6s. 6d.
LOCALISATION OF FAULTS IN ELECTRIC LIGHT MAINS—see page 10.
PHŒNIX FIRE OFFICE RULES FOR ELECTRIC LIGHT INSTALLATIONS AND ELECTRICAL POWER INSTALLATIONS By Musgrave Heaphy, C.E. 6d.
ELECTRICAL INSTALLATIONS FOR ARCHITECTS, BOROUGH SURVEYORS, &c. By F. J. Warden Stephens. 2s. 6d.
ALTERNATING CURRENT WIRING AND DISTRIBUTION. By Wm. Leroy Emmett. 4s. 6d.
THE ELECTRIC WIRING AND FITTING DETAILS BOOK. By W. Perren Maycock. 2s. 6d.
ELECTRIC WIRING, FITTINGS, SWITCHES AND LAMPS. By W. Perren Maycock. 6s.
ARC LAMPS and HOW to MAINTAIN THEM. By Smithson and Sharpe. 1s.
ELECTRICITY IN TOWN AND COUNTRY HOUSES. By Percy E. Scrutton. 2s. 6d.
THE INTERNAL WIRING OF BUILDINGS. By H. M. Leaf. 3s. 6d.

1, 2 and 3, Salisbury Court, Fleet Street, London, E.C.

TELEGRAPHY AND TELEPHONY.

TEMPERATURE COEFFICIENTS OF CONDUCTIVITY COPPER. By Clark, Forde and Taylor—*see page 14.*
STUDENTS' GUIDE TO SUBMARINE CABLE TESTING—*see page 15.*
PRACTICAL NOTES FOR ELECTRICAL STUDENTS—*see page 13.*
ELECTRICITY AND THE ELECTRIC TELEGRAPH. With numerous illustrations. By George B. Prescott. Eighth Edition, 2 vols., 8vo, cloth, £1. 10s.
ELECTRIC TELEGRAPHY. By Prof. E. J. Houston, and A. E. Kennelly. 4s. 6d.
ELECTRICAL ENGINEERING FORMULÆ—*see page 5.*
THE ELECTRIC TELEGRAPH: Its History and Progress. With Descriptions of some of the Apparatus. By R. Sabine, C.E., F.S.A. Limp cloth, 3s.
SUBMARINE CABLE LAYING AND REPAIRING—*See p. 10.* [Smith 21s
RISE AND EXTENSION OF SUBMARINE TELEGRAPHY. By Willoughby
TELEGRAPHY. By W. H. Preece, C.B., F.R.S., M.I.C.E., and Sir J. Sivewright, K.C.M.G., M.A. New Edition, revised and enlarged. 6s. [12s. 6d.
MODERN PRACTICE OF THE ELECTRIC TELEGRAPH. By Frank L. Pope.
HANDBOOK OF THE ELECTRO-MAGNETIC TELEGRAPH. A. E. Loring. 2s.
"THE ELECTRICIAN" PRIMERS—*see page 8.*
ELECTRICAL TESTING FOR TELEGRAPH ENGINEERS—*see page 15.*
TELEGRAPHIC CONNECTIONS: Embracing Recent Methods in Quadruplex Telegraphy. By Charles Thom and W. H. Jones. 7s. 6d.
PRACTICAL GUIDE TO THE TESTING OF INSULATED WIRES AND CABLES. By Herbert Laws Webb. 4s. 6d.
A HANDBOOK OF ELECTRICAL TESTING. By H. R. Kempe. Fifth edition, revised and enlarged. 18s.
SUBMARINE TELEGRAPHS: Their History, Construction and Working. By Charles Bright, F.R.S.E. £3. 3s. net. [6th Edition. 3s. 6d.
ELECTRICITY IN ITS APPLICATION TO TELEGRAPHY. By T. E. Herbert.
THE TELEGRAPHIST'S GUIDE TO THE NEW EXAMINATIONS IN TECHNICAL TELEGRAPHY. By James Bell. Fourth Edition. 1s. 6d.
THE ELECTRIC TELEGRAPH: Being selected Extracts from the letters of the late Sir W. F. Cooke (1836-39) relating to the Invention and Development of the Electric Telegraph. Edited by F. H. Webb, late Secretary Inst. E.E. Cloth, 3s.
SIGNALLING ACROSS SPACE WITHOUT WIRES—*see page 13.*
TÉLÉGRAPHIE SANS FILS. By André Broca. 3s. 6d.
WIRELESS TELEGRAPHY. By Richard Kerr. 1s.
WIRELESS TELEGRAPHY: A History of. By J. J. Fahie. 6s.
CENTENARY OF THE ELECTRIC CURRENT, 1799-1899—*see page 13.*
PRACTICAL INFORMATION FOR TELEPHONISTS. By T. D. Lockwood. 5s.
COMMERCIAL AND DOMESTIC TELEPHONY. By M. Byng and F. Bell. 2s. 6d.
TELEPHONES: Their Construction and Fitting. 4th Edition By F. C. Allsop. 5s.
ELECTRIC TELEPHONY. By Prof. E. J. Houston and A. E. Kennelly, 4s. 6d
TELEPHONE LINES AND THEIR PROPERTIES. By W. J. Hopkins. 2nd Ed. 6s. 6d
THE PRACTICAL TELEPHONE HANDBOOK. By Joseph Poole. 2nd Ed. 5s.
MANUAL OF THE TELEPHONE. By W. H. Preece and A. J. Stubbs. Over 500 pages and 334 illustrations. 15s
THE TELEPHONE SYSTEMS OF THE CONTINENT OF EUROPE. By A. R. Bennett, M.I.E.E. 169 Illustrations. 15s
THE TELEPHONE: Outlines of the Development of Transmitters and Receivers. By Prof. W. J. Hopkins. 3s. 6d.
AMERICAN TELEPHONE PRACTICE. By Kempster B. Miller. 9s.
THE TELEPHONE, THE MICROPHONE, AND THE PHONOGRAPH. By Count du Moncel. Third Edition. 5s.
THE MAGNETO HAND TELEPHONE. By Norman Hughes. 3s. 6d.
PHILIPP REIS, INVENTOR OF THE TELEPHONE. A Biographical Sketch. By Silvanus P. Thompson, B.A., D.Sc. With portrait and wood engravings. 8vo, cloth, 7s. 6d.
ELECTROMAGNETIC THEORY—*see page 4.*
TRAITÉ DE TÉLÉGRAPHIE ELECTRIQUE. By H. Thomas. £1.
TRAITÉ DE TÉLÉGRAPHIE SOUS-MARINE. By E. Wünschendorff 13s.
LA TELEPHONIE. By E. Piérard. 8s. 6d.

ELECTRO-CHEMISTRY & ELECTRO-METALLURGY.

THEORY AND PRACTICE OF ELECTROLYTIC METHODS OF ANALYSIS.
By Dr. B. Neumann. Translated by J. B. C. Kershaw. 10s. 6d.
"THE ELECTRICIAN" PRIMERS—*see page 8.*
ELECTRIC SMELTING AND REFINING. A Translation of the Second Edition of Dr. W. Börcher's "Elektro-Metallurgie." By W. G. McMillan. 21s.
ELECTRO-PLATING: A Practical Handbook on the Deposition of Copper, Silver, Nickel, Gold, Aluminium, Brass, Platinum, &c. &c. By John W. Urquhart. Third Edition, 5s.
ELECTRO-DEPOSITION: A Practical Treatise on the Electrolysis of Gold, Silver, Copper, Nickel, and other Metals and Alloys. By Alexander Watt, F.R.S.S.A. With numerous Illustrations. Third Edition. Crown 8vo, cloth, 9s.
PRACTICAL METALLURGY AND ASSAYING. By A. H. Hiorns. 6s.
ELECTRO METALLURGY: Practically treated. By Alexander Watt, F.R.S.A. Ninth Edition. 12mo, cloth boards, 4s.
ART OF ELECTRO-METALLURGY, INCLUDING ALL KNOWN PROCESSES OF ELECTRO-DEPOSITION. By G. Gore, LL.D., F.R.S. With 56 Illustrations, fcp. 8vo, 6s.
THE ART OF ELECTROLYTIC SEPARATION OF METALS—*see page 11.*
ELECTRO-METALLURGY, TREATISE ON. By Walter G. M'Millan. 10s. 6d.
ELECTRO-PLATER'S HANDBOOK. By G. E. Bonney. Third Edition. 3s.
ELECTRO-CHEMISTRY—*see page 11.*
ELECTRO-DEPOSITION OF METALS. By Dr. G. Langbein. 25s.
CHEMISTRY FOR ENGINEERS AND MANUFACTURERS, By Bertram Blount, F.I.C., F.C.S., and A. G. Bloxam. In two vols. Vol. I.—Chemistry of Engineering, Building and Metallurgy, 10s. 6d. Vol. II., Chemistry of Manufacturing Processes, 16s.
ELEMENTS OF ELECTRO-CHEMISTRY. Treated Experimentally. By Dr. Lüpke. Translated from the German by W. M. Pattison Muir. 7s. 6d.
ELEMENTS OF ELECTRO-CHEMISTRY. By Prof. M. le Blanc. Translated by W. R. Whitney. 6s.
QUANTITATIVE CHEMICAL ANALYSIS BY ELECTROLYSIS. Translated from the German of Dr. A. Classen and Dr. W. Löb by W H. Herrick and B. B. Boltwood. 12s. 6d.
ELECTRICAL ENGINEERING FORMULÆ—*see page 5.*
TRAITE THÉORIQUE ET PRATIQUE D'ÉLECTROCHIMIE. By Donato Tommasi, Docteur-ès-Sciences. Large 8vo, 1,200 pages. 30s.
LE FOUR ELECTRIQUE. By M. Henri Moissan. 12s. 6d.
ELEKTRO-METALLURGIE. By Dr. W. Börchers. 6s. 6d.
JAHRBUCH DER ELEKTROCHEMIE. By Prof. Nernst and Dr. Börchers. 16s.

MEDICAL ELECTRICITY.

LECTURES ON PHYSIOLOGY. First Series on Animal Electricity. By Augustus D. Waller, F.R.S. 5s.
ELECTRO-PHYSIOLOGY. By Prof. W. Biedermann. Translated by Frances A. Welby. Vols. I. and II., each 17s. net.
PRACTICAL ELECTRICITY IN MEDICINE AND SURGERY. By Drs. G. A. Liebig and G. H. Rohé. Royal 8vo, 400 pages, profusely illustrated, 11s. 6d.
ELECTRICITY IN THE DISEASES OF WOMEN. By G. Betton Massey, M.D. 2nd Ed., 12mo, 8s. 6d. [by A. de Watteville, M.D., &c. Demy 8vo, 18s.
MANUAL OF ELECTRO-THERAPEUTICS. By W. Erb, M.D. Translated
INTERNATIONAL SYSTEM OF ELECTRO-THERAPEUTICS. For Students, General Practitioners, and Specialists. Edited by H. R. Bigelow. 34s.
CURRENT FROM THE MAIN: The Medical Employment of Electric Lighting Currents. By W. S. Hedley, M.D. Second Edition. 2s. 6d.
THERAPEUTIC ELECTRICITY. By W. S. Hedley, M.D. 8s. 6d.
PRACTICAL MUSCLE TESTING AND THE TREATMENT OF MUSCULAR Atrophies. By W. S. Hedley, M.D. 3s. 6d. [Cloth, 5s.
USES OF ELECTROLYSIS IN SURGERY. By Dr W. E. Steavenson.
ELECTRICITY AND ITS MANNER OF WORKING IN THE TREATMENT OF DISEASE. A Thesis for the M.D. Cantab Degree, 1884. By the late William E. Steavenson, M.D., M.R.C.P. To which is appended an Inaugural Medical Dissertation on Electricity for the M.D. (Edin.) Degree, written in Latin by Robert Steavenson, M.D., in 1778, with Translation. 4s. 6d.
ELECTRICITY IN ELECTRO-THERAPEUTICS. By Prof. E. J. Houston, Ph.D., and A. E. Kennelly, D.Sc. Cloth, 4s. 6d.
THE ELECTRO-MAGNET AND ITS EMPLOYMENT IN OPHTHALMIC SURGERY. By Simeon Snell, M.R.C.S. 3s. 6d.
VALUE OF ELECTRICAL TREATMENT. By Julius Althaus M.D. 3s. 6d.

ELECTRICAL INSTRUMENTS, BELLS, &c.

"THE ELECTRICIAN" PRIMERS—*see page 8.*
ELECTRICAL INSTRUMENT MAKING FOR AMATEURS. A Practical Handbook. By S. R. Bottone. Fifth Edition. 3s.
INDUCTION COILS AND COIL MAKING. By F. C. Allsop. 3s. 6d.
THE POTENTIOMETER AND ITS ADJUNCTS—*See page 10.*
INDUCTION COILS: A Practical Manual for Amateur Coil Makers. By G. E. Bonney. 3s.
ELECTRICAL ENGINEERING FORMULÆ—*see page 5.*
DYNAMOMETERS AND THE MEASUREMENT OF POWER. By Prof. J. J. Flather. 8s. 6d.
RUHMKORFF INDUCTION COILS. By H. S. Norrie. 2s. 6d.
PRACTICAL ELECTRIC BELL FITTING. By F. C. Allsop. 3s. 6d.
INTENSITY COILS; HOW MADE AND HOW USED. By Dyer. 1s. 2d.
THE INDUCTION COIL IN PRACTICAL WORK. By Lewis Wright, 4s.
THE BELL-HANGER'S HANDBOOK. By F. B. Badt. 4s. 6d.
ELECTRIC BELL CONSTRUCTION. By F. C. Allsop. Fifth Edition. 3s. 6d.
ELECTRIC BELLS, AND ALL ABOUT THEM. By S. R. Bottone. 4th edition. 3s.
ELECTRICAL APPARATUS FOR AMATEURS. By Various Authors. 1s.
THE PHONOGRAPH: and How to Construct It. By W. Gillett. 5s.
INSTRUMENTS ET METHODES DE MESURES ELECTRIQUES INDUSTRIELLES. By H. W. Armagnat. 10s. 6d.

RONTGEN RAY LITERATURE.

PRACTICAL RADIOGRAPHY. By A. W. Isenthal and H. S. Ward. 2nd Ed. 2s. 6d.
THE A B C OF THE X-RAYS. By W. H. Meadowcroft. 4s.
THE X-RAY; or, Photography of the Invisible. By W. J. Morton and E. W. Hammer. 4s.
BIBLIOGRAPHY OF X-RAY LITERATURE AND RESEARCH—*see p. 13.*
THE NEW PHOTOGRAPHY. By A. B. Chatwood. 4s.
PROF. RÖNTGEN'S "X"-RAYS AND THEIR APPLICATIONS TO THE New Photography. By August Dittmar. 1s
X-RAYS, THEIR PRODUCTION & APPLICATION. By F. S. Kolle, M.D. 4s. 6d.
THE DISCHARGE OF ELECTRICITY THROUGH GASES. By J. J. Thomson, D.Sc., F.R.S. 4s. 6d. nett.
THE ARCHIVES OF THE RÖNTGEN RAY. Edited by T. Moore, F.R.C.S., and E. Payne, M.A. Price 4s.; Ann. Sub. 16s. An illustrated quarterly publication containing the proceedings of the Röntgen Society of London, besides many original articles and illustrations.
RADIATION: An Elementary Treatise on Electromagnetic Radiation and on Röntgen and Cathode Rays. By H. H. Francis Hyndman, B Sc. 6s.
LIGHT, VISIBLE AND INVISIBLE. By Prof. Silvanus P. Thompson, D.Sc., F R.S. 6s. nett.
MANUAL OF MEDICAL ELECTRICITY, WITH CHAPTERS ON THE Röntgen Rays. By Dawson Turner, B.A., M.D., &c. Price 7s. 6d.
RÖNTGEN RAYS IN MEDICAL WORK. By D. Walsh, M.D. 2nd Ed. 10s. 6d. net.
RADIOGRAPHY AND THE X-RAYS IN PRACTICE AND THEORY. By S. R. Bottone. 3s.
LES RAYONS-X. By C. E. Guilleaume. 2s. 6d.
RONTGEN RAYS, AND THE PHENOMENA OF THE CATHODE AND ANODE. By E. P. Thompson and W. A. Anthony. 7s. 6d.
LA RADIOGRAPHIE APPLIQUEE A L'ETUDE DES ARTHROPATHIES DEFORMANTES. By Dr. F. Barjon. Price 6s. 6d.
LA TECHNIQUE DES RAYONS X. By A. Herbert. Price 4s. 6d.
LUCE E RAGGI RÖNTGEN. By Professor O. Murani, with preface by Prof. R. Ferrini. 7s. 6d.

PHOTOGRAPHY, &c.

SCIENCE AND PRACTICE OF PHOTOGRAPHY. By Chapman Jones, F.C.S. New Edition, 3s. Cloth, 4s.
PHOTOGRAPHY WITH EMULSIONS. By Capt. Sir W. de W. Abney, F.R.S. 3s.

BOOK OF THE LANTERN. By T. C. Hepworth, F.C.S. New Edition. 4s.
PRACTICAL POCKET BOOK OF PHOTOGRAPHY. By Dr. E. Vogel.
Translated by E. C. Conrad, F.C.S. 1s. 3d. Cloth, 1s. 9d.
CHEMISTRY OF PHOTOGRAPHY. By Prof. R. Meldola. Crown 8vo. 6s.
FIRST PRINCIPLES OF PHOTOGRAPHY. By C. J. Leaper, F.C.S. 5s
PHOTOGRAPHY FOR AMATEURS. By T. C. Hepworth. Cloth, 1s. 6d.
PHOTOGRAPHIC AND OPTICAL ELECTRIC LAMPS. By R. Kennedy. 2s. 6d.
INSTRUCTION IN PHOTOGRAPHY. By Capt. W. de W. Abney, F.R.S.
Ninth Edition. 4s.
OPTICS OF PHOTOGRAPHY. By B. J. Trail-Taylor. 3s. 6d.
PHOTOGRAPHIC REFERENCE BOOK. By W. A. Watts. 6s. 6d.

DICTIONARIES, DIRECTORIES, TABLES, &c.

"**THE ELECTRICIAN**" **ELECTRICAL TRADES' DIRECTORY AND**
HANDBOOK—see page 3.
WILLING'S BRITISH AND IRISH PRESS GUIDE. A concise and comprehensive Index to the Press of the United Kingdom, with Lists of principal Colonial and Foreign Journals. 1s.
HAZELL'S ANNUAL. A Cyclopædic Record of Men and Topics of the Day. 3s. 6d.
STATESMAN'S YEAR BOOK: A Statistical and Historical Annual of the
States of the World. Revised after Official Returns. 10s. 6d.
WHITAKER'S ALMANACK. 2s. 6d.
SELL'S DIRECTORY OF REGISTERED TELEGRAPHIC ADDRESSES:
From Official Lists. Price, including supplements, subscribers, 18s., non-subscribers, 21s.
KELLY'S POST OFFICE AND OTHER DIRECTORIES. (Particulars and lowest prices on application.)
GAS AND ELECTRIC LIGHTING COMPANIES' DIRECTORY AND
STATISTICS, &c. Revised to June, 1899. Twentienth issue. 6s.
ELECTRICAL ENGINEERING FORMULÆ—see page 5.
A POCKET-BOOK OF ELECTRICAL RULES AND TABLES, for the use
of Electricians and Engineers. By John Munro, C.E., and Andrew Jamieson, C.E., F.R.S.E. Thirteenth Edition. Revised and enlarged. Pocket size, leather, 8s. 6d., post free.
USEFUL RULES AND TABLES FOR ENGINEERS AND OTHERS. By
Prof. Rankine. With Appendix by Andrew Jamieson, C.E. 10s. 6d.
A MANUAL OF RULES, TABLES AND DATA FOR MECHANICAL
ENGINEERS. By D. K. Clark. Cloth, 16s.; half-bound 20s.
LAXTON'S PRICE BOOK. Published Annually. 4s.
LOCKWOOD'S BUILDERS' PRICE BOOK. Edited by Francis T. W. Miller. 4s.
THE ELECTRICAL ENGINEER'S POCKET-BOOK OF MODERN RULES,
FORMULÆ, TABLES, AND DATA. By H. R. Kempe, M.I.E.E. Price 6s. post free.
ENGINEER'S YEAR BOOK of Formulæ, Rules, Tables, Data, and Memoranda.
By H. R. Kempe. 8s.
THE STANDARD ELECTRICAL DICTIONARY. By Dr. T. O'Conner
Sloane. Second Edition. 7s.
DICTIONARY OF CHEMISTRY (Watts'). Revised and rewritten by M. M. Pattison Muir, M.A., and H. F. Morley, M.A., D.Sc. Vols. I. and II, 42s. each; Vol. III., 50s; Vol. IV., 63s.
A DICTIONARY OF APPLIED CHEMISTRY. By T. E. Thorpe, assisted
by Eminent Contributors. Vols. I. and II., £2. 2s. each; Vol. III., £2. 3s.
THE GRAPHIC ATLAS AND GAZETTEER OF THE WORLD. 128 Map
Plates and Gazetteer of 280 pages. Crown 4to, cloth, 12s. 6d; half-morocco, 15s.; full morocco, 21s.
TABLES OF EQUIVALENTS OF UNITS OF MEASUREMENT. By Carl
Hering. 2s. 6d.
DICTIONARY OF METRIC AND OTHER USEFUL MEASURES. By
Latimer Clark, F.R.S., &c. Crown 8vo 6s
METRICAL TABLES. By Sir G. L. Molesworth. Third edition. 2s.
BLACKIE'S MODERN CYCLOPÆDIA OF UNIVERSAL INFORMATION.
Edited by Charles Annandale, M.A., LL.D. Eight vols., cloth, 6s. each; half-morocco, 8s. 6d. each.
HOUSTON'S DICTIONARY OF ELECTRICAL WORDS, TERMS,
AND PHRASES. Fourth Edition, greatly enlarged. 30s., post free.
A POCKET DICTIONARY OF ELECTRICAL WORDS, TERMS AND
PHRASES. By E. J. Houston. Fourth Edition. 10s. nett.

MANUAL OF ELECTRICAL UNDERTAKINGS.—*See page 16.*
THE UNICODE. Desk Edition (demy 8vo) or Pocket Edition (5in. by 2¾in.) 2s. 6d.
PRACTICAL DICTIONARY OF MECHANICS. By Edward H. Knight.
  Three volumes. With 15,000 illustrations. £3. 3s. Supplementary volume, £1. 1.
DIRECTORY OF DIRECTORS: A list of Directors of Public Companies, with
  the concerns with which they are connected. 15s.
STOCK EXCHANGE YEAR-BOOK: A Digest of Information relating to the
  Origin, History, and Present Position of the Public Securities and Joint Stock Companies known to the
  Markets of the United Kingdom. 25s.
UNIVERSAL ELECTRICAL DIRECTORY. 6s.
ENCYCLOPÆDIA BRITANNICA. Ninth Edition in 24 vols. Cloth extra 30s.
  per vol., half Morocco or half Russia, 36s. per vol. Part (4 in vol.) containing any special article, 7s. 6d.
TABLES AND FORMULÆ FOR ELECTRIC STREET RAILWAY EN-
  GINEERS. Compiled by E. A. Merrill. 4s. 6d.
BARLOW'S TABLES OF SQUARES, CUBES, SQUARE ROOTS, CUBE
  ROOTS. Reciprocals of all Integer Numbers up to 10,000. Post 8vo, cloth, 6s.
A B C FIVE-FIGURE LOGARITHMS. By C. J. Woodward, B.Sc. Cloth, 4s.
ILLUSTRATIONS OF THE C.G.S. SYSTEM OF UNITS. With Tables of
  Physical Constants. By Prof. J. D. Everett. 5s.
LOGARITHMIC AND TRIGONOMETRICAL TABLES FOR APPROXI-
  MATE CALCULATION, &c. By J. T. Bottomley, M.A. 1s.
MATHEMATICAL TABLES. By James Pryde, F.E.I.S. 4s. 6d.
SPONS' ENGINEERS' DIARY. Published annually. 3s. 6d.
SPONS' ENGINEERS' PRICE-BOOK. 7s. 6d.
ANALYSTS' LABORATORY COMPANION. By A. Johnson. Second Edition. 5s.
ELECTRICAL TABLES AND MEMORANDA. By Silvanus P. Thompson,
  D.Sc., and Eustace Thomas. Waistcoat-pocket size, 1s.; post free, 1s. 1d.
DIGEST OF THE LAW OF ELECTRIC LIGHTING, &c.—*see page 8.*
CHAMBERS' ENCYCLOPÆDIA. Ten volumes. Cloth, £5; half morocco
  or half calf, £7. 10s.; and in better bindings up to £15.
A POCKET-BOOK OF USEFUL FORMULÆ AND MEMORANDA FOR CIVIL
  AND MECHANICAL ENGINEERS. By Sir G. L. Molesworth, K.C.I.E., and R. B. Molesworth, M.A.
  Twenty-fourth Edition. 6s.
WHITTAKER'S MECHANICAL ENGINEER'S POCKET BOOK. By P. R.
  Bjorling. 5s.
THE MECHANICAL ENGINEER'S POCKET-BOOK. By D. K. Clark. Third
  Edition. 6s.
TABLES, MEMORANDA, AND CALCULATED RESULTS, for Mechanics,
  Engineers, &c. By Fr. Smith. Fifth Edition, revised and enlarged. Waistcoat pocket size, 1s. 6d.
PRACTICAL ENGINEER'S HANDBOOK. By W. S. Hutton. 18s.
REID'S PATENT INDEXED READY RECKONER. With fore-edge Index.
  Third Edition. 2s. 6d.
ANNUAIRE POUR L'AN 1900. 1s. 6d.
GUIDE TO OUR OLD HOMES ACROSS THE WATER. By Oskar Moll. 5s.

MISCELLANEOUS.

WIRE: ITS MANUFACTURE AND USES. By J. Bucknall Smith. 7s. 6d.
A TREATISE ON BESSEL FUNCTIONS AND THEIR APPLICATIONS TO
  PHYSICS. By Prof. A. Gray, M.A., F.R.S.E., and G. B. Mathews, M.A. 16s.
NATURAL PHILOSOPHY: An Elementary Treatise. By Prof. A. Privat
  Deschanel. Translated and added to by Prof. J. D. Everett, D.C.L., F.R.S. Twelfth Edition. Cloth, 18s.
  Also in four parts.—I. MECHANICS, HYDROSTATICS, &c.; II. HEAT; III. ELECTRICITY AND MAGNETISM
  IV. SOUND AND LIGHT. 4s. 6d. each
LIGHTING BY ACETYLENE: Generators, Burners and Electric Furnaces. By
  W. E. Gibbs. 7s. 6d.
MATHEMATICAL FORMULÆ, INTERPRETATION OF. By E. J. Houston
  and A. E. Kennelly. 3s. net
GRAPHICAL CALCULUS. By A. H. Barker. 4s. 6d.
LIGHTNING CONDUCTORS AND LIGHTNING GUARDS. By Prof.
  Oliver J. Lodge, D.Sc., &c. With numerous illustrations. 15s.
ELEMENTARY TEXT-BOOK OF PHYSICS. By Prof. Everett, D.C.L.,

THE STEAM ENGINE INDICATOR, AND INDICATOR DIAGRAMS—
see page 15.
ELECTRICAL EXPERIMENTS. By G. E. Bonney 2s. 6d.
ELECTRICAL ENGINEERING AS A PROFESSION, AND HOW TO ENTER IT. Official Edition. By A. D. Southam. 3s. 6d.
AID BOOK TO ENGINEERING ENTERPRISE. By Ewing Matheson, M.I.C.E. New edition. 850 pages. £1. 1s.
ENGINEERING CONSTRUCTION IN IRON, STEEL AND TIMBER. By W. H. Warren. 16s. net.
RAILWAY SIGNALLING. By H. R. Wilson. Royal 4to, cloth, 7s 6d., post free, 8s.
PRACTICAL PHYSICS. By R. T. Glazebrook and W. N. Shaw. 7s. 6d.
EXAMPLES IN PHYSICS. By D. E. Jones, B Sc. Fcap. 8vo, 3s. 6d.
JAMES CLERK MAXWELL AND MODERN PHYSICS. By R. T. Glazebrook, M.A. 3s. 6d.
EXPERIMENTAL SCIENCE. By G. M. Hopkins. 680 engravings. Cloth, 15s.
PRACTICAL CHEMISTRY, INCLUDING ANALYSIS. By J. E. Bowman and Prof. C. L. Bloxam. Eighth Edition. With 90 engravings, fcap. 8vo, 5s. 6d.
OUTLINES OF GENERAL CHEMISTRY. By Prof. W. Ostwald. Translated by Dr. J. Walker. 10s. net.
LESSONS IN ELEMENTARY CHEMISTRY. By Sir H. E. Roscoe, F.R.S., Sixth Edition, 4s. 6d.
A TREATISE ON CHEMISTRY. By Sir H. E. Roscoe, F.R.S., and C. Schorlemmer. Vols. I. and II.: Inorganic Chemistry. Vol. I., The Non-Metallic Elements, 21s. Vol. II., Metals, in two parts, 18s. each. Vol. III.: Organic Chemistry. Parts 1, 2, 4, and 6, 21s. each.; Parts 3 and 5, 18s. each.
THE PRINCIPLES OF CHEMISTRY. By Prof. D. Mendeléeff. Translated by George Kamensky. Two Volumes. 36s.
BLOXAM'S INORGANIC AND ORGANIC CHEMISTRY, with Experiments. Re-written and Revised by Prof. J. M. Thomson, F.R.S., and Arthur G. Bloxam. Eighth Edition. 18s. 6d.
PRACTICAL CHEMISTRY: Qualitative and Quantitative Analysis. By William G. Valentin, F.C.S. Ninth Edition, by Dr. W. R. Hodgkinson, F.R.S. 9s.
A LABORATORY COURSE OF EXPERIMENTAL PHYSICS. By W. J. Loudon, B.A., and J. C. McLennan, B.A. 8s. 6d., nett.
AN ELEMENTARY TREATISE ON THE CALCULUS FOR ENGINEERING STUDENTS. By John Graham, B.A., B.E. Crown 8vo. Cloth, 7s. 6d.
MISCELLANEOUS PAPERS. By the late Prof. Heinrich Hertz. Translated by D. E. Jones, B.Sc., and G. A. Schott, B.A., B.Sc. With an Introduction by Prof. Philip Lenard. 10s. 6d., nett
THE STEAM ENGINE CONSIDERED AS A THERMO-DYNAMIC MACHINE. By J. H. Cotterill, M.A., F.R.S. Third Edition. 15s.
ELEMENTARY PRACTICAL CHEMISTRY AND QUALITATIVE ANALYSIS FOR ORGANISED SCIENCE SCHOOLS. By Prof. F. Clowes, D.Sc., and J. Bernard Coleman. 8vo. 3s. 6d.
THE CONSTRUCTOR. A Hand-book of Machine Design. By Prof. F. Reuleaux. Translated from the Fourth German Edition by H. H. Suplee, B.Sc. £1. 11s. 6d.
TABULATED WEIGHTS OF IRON AND STEEL. By C. H. Jordan, M.I.N.A. Fifth Edition. 600 pages, Royal 32mo. 7s 6d.
INORGANIC CHEMISTRY. By Prof. Edward Frankland, D.C.L., F.R.S., and Prof. F. R. Japp, Ph.D., F.R.S. 24s.
A SYSTEM OF INORGANIC CHEMISTRY. By Prof. W. Ramsay, F.R.S. 15s.
METAL TURNING. By a Foreman Pattern-Maker. 4s.
"THE ELECTRICIAN" PRIMERS—*see page 8.*
ALUMINIUM: Its History, Occurrence, Properties, Metallurgy, and Applications, including its Alloys. By J. W. Richards. Second Edition. 511 pages, £1. 1s.
RAILWAY ENGINEERING—Mechanical and Electrical. By J. W. Haldane. 15s.
LIGHT RAILWAYS FOR THE UNITED KINGDOM, INDIA AND THE COLONIES. By J. C. Mackay. 15s.
STREET RAILWAY ROAD BEDS. By Mason D. Pratt and C. A. Alden. 9s.
STREET RAILWAY INVESTMENTS, A STUDY IN VALUES. By E. E. Higgins. 9s.
THE LAW SPECIALLY RELATING TO TRAMWAYS AND LIGHT RAILWAYS, with the Tramways Act, 1870, the Light Railways Act, 1896, &c. By Seward Brice Q C. 12s. 6d
LAW AND PRACTICE RELATING TO LETTERS PATENT FOR INVENTIONS. By Robert Frost, B.Sc. Second Edition. 30s.
TRAMWAYS: Their Construction and Working. By D. K. Clark. 28s.
AMERICAN STREET RAILWAY INVESTMENTS. Published annually 22s. 6d!
PRINCIPLES AND PRACTICE OF THE VALUATION OF GAS, ELEC

CENTRAL STATION BOOK-KEEPING. By H. A. Foster. 10s. 6d.
CENTRAL STATION MANAGEMENT AND FINANCE. By H. A. Foster. 7s. 6d.
PRACTICAL TREATISE ON THE STEAM-ENGINE. By Arthur Rigg. £1. 5s.
THERMODYNAMICS OF THE STEAM ENGINE AND OTHER HEAT ENGINES. By Prof. C. H. Peabody. 21s.
AN ELEMENTARY TREATISE ON STEAM. By Prof. John Perry. 4s. 6d.
STEAM ENGINES AND OTHER HEAT ENGINES. By Prof. J. A. Ewing, M.A., F.R.S. 15s.
RECENT IMPROVEMENTS IN THE STEAM ENGINE. By John Bourne. 6s.
A HANDBOOK OF THE STEAM ENGINE. By John Bourne, C.E. 9s.
STEAM AND THE STEAM-ENGINE. By Prof. Andrew Jamieson, M.Inst.C.E. Text-Book, 8s. 6d.; Elementary Manual, 3s. 6d.
THE STEAM ENGINE: A Treatise on Steam-Engines and Boilers. By D. K. Clark. Two vols. 50s. net.
MARINE BOILERS: Their Construction and Working, Dealing more Especially with Tubular Boilers. By L. E. Bertin. Trans. by Leslie S. Robinson. Preface by Sir Wm. White, K.C.B. 18s.
STEAM BOILERS. By G. Halliday. 7s. 6d.
TEXT-BOOK ON THE STEAM ENGINE. By Prof. T. M. Goodeve, M.A. 11th Ed. 6s
THE STEAM ENGINE AND GAS AND OIL ENGINES. By Prof. John Perry, F.R.S. 7s. 6d. nett.
GAS AND PETROLEUM ENGINES. By Prof. William Robinson. New Edition. Entirely Re-written. Ready shortly.
ON GAS ENGINES. By Prof. T. M. Goodeve, M.A. 2s. 6d.
GAS, OIL AND AIR ENGINES. By Bryan Donkin, Second Edition. 25s.
GAS AND OIL ENGINES. By Dugald Clerk. 6th Ed., revised & enlarged. 8vo. 15s.
GAS AND PETROLEUM ENGINES. Translated from the French of H. de Graffigny. By A. G. Elliott, B.Sc. 2s. 6d.
TREATISE ON THE "OTTO" CYCLE GAS ENGINE. By Wm. Norris, M.I.M.E. 207 Illustrations. 10s. 6d.
THE CYCLES OF GAS AND OIL ENGINES. By J. D. Roots. 4s. 6d.
THEORY OF THE GAS ENGINE. By Dugald Clerk. 2s.
HEAT AND HEAT ENGINES. By W. C. Popplewell, M.Sc. 6s.
HEAT. By Prof. P. G. Tait. 6s.
THE THEORY OF HEAT. By J. Clerk Maxwell, M.A. Edited by Lord Rayleigh. 4/6
AN ELEMENTARY TREATISE ON HEAT. By William Garnett, M.A., D.C.L. Fifth Edition, 6s.
HEAT A MODE OF MOTION By John Tyndall, D.C.L. 12s.
MOTIVE POWER AND GEARING—*see page 9*.
A TREATISE ON ELEMENTARY DYNAMICS. By William Garnett, M.A., D.C.L. Fifth Edition, 6s.
MODERN MECHANISM, exhibiting the latest Progress in Machines, Motors and the Transmission of Power. Edited by Park Benjamin, LL.B., Ph.D. 30s.
ELEMENTARY MECHANICS, INCLUDING HYDROSTATICS AND PNEUMATICS. By Oliver J. Lodge, D.Sc. Lond. New Edition. 4s. 6d.
APPLICATIONS OF DYNAMICS TO PHYSICS AND CHEMISTRY. By Prof. J. J. Thomson. 7s. 6d.
MECHANICS APPLIED TO ENGINEERING. By Prof. John Goodman. 7s. 6d.
MANUAL OF PHYSICO-CHEMICAL MEASUREMENTS. By Prof. W. Ostwald. Translated by James Walker, D.Sc. 7s. nett. [12s. 6d. net.
TEXT-BOOK OF MECHANICAL ENGINEERING. By W. J. Lineham. 3rd Ed.
THE CONSERVATION OF ENERGY. By Balfour Stewart. 7th Edition. 5s.
AN ELEMENTARY TEXT-BOOK OF APPLIED MECHANICS. By David Allan Low. 2s.
PRACTICAL MECHANICS AND MACHINE DESIGN. NUMERICAL EXAMPLES IN. By R. G. Blaine. 2s. 6d.
LESSONS IN APPLIED MECHANICS. By Prof. J. H. Cotterill and J. H. Slade. 5s. 6d.
THE COMMERCIAL MANAGEMENT OF ENGINEERING WORKS. By F. G. Burton. 12s. 6d.
COMMERCIAL ORGANISATION OF FACTORIES. By J. Slater Lewis. 28s.
THE COMMERCIAL AND BUSINESS ASPECTS OF MUNICIPAL ELECTRICITY SUPPLY. By Alfred H. Gibbings. 15s.

THE MANAGEMENT OF SMALL ENGINEERING WORKSHOPS. By Arthur H. Barker, B.A., B.Sc. 7s. 6d.
APPLIED MECHANICS. By Prof. J. Perry, D.Sc., F.R.S. 7s. 6d.
AUTO CARS. By D. Farman. Trans. from the French by L. Serriallier. Cr. 8vo., 5s.
THE AUTOMOTOR AND HORSELESS VEHICLE POCKET BOOK OF AUTOMOTIVE FORMULÆ AND COMMERCIAL INTELLIGENCE. 1s. 6d.
MOTOR CARS: Or Power Carriages for Common Roads. By A. J. Wallis-Tayler, A.M.I.C.E. Crown 8vo. 4s. 6d.
MECHANICAL ROAD CARRIAGES: A Course of Cantor Lectures. By. W. Worby Beaumont, M.I.C.E. 1s.
THE MECHANICS OF MACHINERY. By Prof. A. B. W. Kennedy. 8s. 6d.
MACHINE DRAWING AND DESIGN FOR ENGINEERING STUDENTS. By Prof. William Ripper. 2nd Edition. 12s. 6d., net.
MACHINE DRAWING AND DESIGN, AN INTRODUCTION TO. By David Allan Low (Whitworth Scholar). With 65 illustrations and diagrams, crown 8vo, 2s.
MACHINE DRAWING AND DESIGN, MANUAL OF. By David Allan Low and Alfred Wm. Bevis. 7s. 6d.
THE CALCULUS FOR ENGINEERS. By Prof. John Perry, F.R.S. 7s. 6d.
DIGEST OF THE BRITISH AND FOREIGN PATENT LAWS. By A. J. Boult, M.I.M.E. 5s.
APPLIED MECHANICS. By Prof. J. W. Cotterill. Third Edition. 18s.
MACHINE DRAWING. By G. Halliday. First Course. Part II. ELECTRICAL ENGINEERING, 2s. 6d Second Course, 6s.
ELEMENTS OF MACHINE DESIGN. By Prof. W. C. Unwin. Part I., General Principles, 6s. Part II., Chiefly on Engine Details, 4s. 6d.
ENGINEER'S SKETCH BOOK of Mechanical Movements, Devices, Appliances. Contrivances used in Design and Construction of Machinery for every purpose. Nearly 2,000 illus. 10s. 6d.
ACETYLENE, LIGHTING BY: Generators, Burners and Electric Furnaces. By W. E. Gibbs. 7s. 6d.
SCIENTIFIC AND TECHNICAL PAPERS of WERNER VON SIEMENS. Translated from the German by E. F. Bamber. Vol. I., Scientific Papers and Addresses. Vol. II., Applied Science. Second Edition. 14s. each.
POPULAR LECTURES AND ADDRESSES ON VARIOUS SUBJECTS IN PHYSICAL SCIENCE. By Sir Wm. Thomson (Lord Kelvin). 3 vols. Vols. I. and III. ready, 7s. 6d. each.
JOURNAL OF THE INSTITUTION OF ELECTRICAL ENGINEERS. Issued at irregular intervals.
ELECTRICAL ENGINEERING FORMULÆ—see page 5.
THE LIFE STORY OF THE LATE SIR CHARLES TILSTON BRIGHT. By E. B. and Charles Bright. Two Vols. 63s.
FORTY YEARS AT THE POST OFFICE, 1850-1890. By F. E. Baines, C.B. Two Vols., 21s
ROMANCE OF ELECTRICITY. By John Munro. Crown 8vo 5s.
POPULAR LECTURES ON SCIENTIFIC SUBJECTS. By Prof. H. von Helmholtz. Two Volumes. 7s. 6d. each.
HYDRAULIC MOTORS: Turbines and Pressure Engines. By G. R. Bodmer, Assoc.M.Inst.C.E. Second Edition. With 179 illustrations. 14s.
WATER OR HYDRAULIC MOTORS. By Philip R. Björling. 208 illustrations. 9s.
PUMPS AND PUMP MOTORS: A Manual for the Use of Hydraulic Engineers. By Philip R. Bjorling. Two Volumes. £3 3s.
CENTRIFUGAL PUMPS, TURBINES AND WATER MOTORS, By C. H. Innes, M.A. 3s. 6d.
HYDRAULIC MACHINERY. By R G. Blaine, M.E. 14s.
SCIENCE LECTURES AT SOUTH KENSINGTON. Containing Lectures by Celebrated Scientists. Two Volumes. 6s. each.
PIONEERS OF SCIENCE. By Dr. O. J. Lodge. 7s. 6d.
PROF CLERK MAXWELL, A LIFE OF. By Prof. L. Campbell, M.A., and W. Garnett, M.A. Second Edition. 7s. 6d.
MICHAEL FARADAY: His Life and Work. By Prof. Silvanus P. Thompson. 5s.
SPINNING TOPS. By Prof. J. Perry. 2s. 6d.
SOAP BUBBLES, and the Forces which Mould Them. By C. V. Boys. 2s. 6d.
THEORY OF LIGHT. By Tolver Preston. 15s. net.
THEORY OF SOUND. By Lord Rayleigh. Vols. I. and II., 12s. 6d. each.
SCREW THREADS. By Paul N. Hasluck. 1s. 6d. (Vol. III. in preparation.
THE INDICATOR HANDBOOK. By C. N. Pickworth. Pt. I. 3s.

THE OLDEST AND BEST WEEKLY ELECTRICAL JOURNAL.

Every Friday, price 6d., post free 6½d.

THE ELECTRICIAN.

"THE ELECTRICIAN" (first series) was established as a weekly newspaper in 1861, in a form exactly similar to the present publication, and after occupying a leading position amongst scientific journals for some years, was discontinued, not, however, before it had become apparent that such a publication occupied an essential place amongst the scientific and technical journals.

From 1865 to 1878 was a period during which the absence of "THE ELECTRICIAN" was much felt, no weekly paper efficiently serving the interests of the electrical profession being then in existence; and, as has frequently been the case, it was left to gentlemen taking a keen interest in the applications of electricity to telegraphic purposes, which were at that time the principal electrical industries, to resuscitate "THE ELECTRICIAN," and once again establish it as the recognised leading organ of electrical science and industry. It will thus be seen that "THE ELECTRICIAN" is the oldest electrical *journal* published. Since its second advent "THE ELECTRICIAN" has made rapid progress, and has continued to record in a full and thorough manner all the great discoveries and experiments in electrical science, and to illustrate how these could be commercially applied and profitably worked. In the volumes of "THE ELECTRICIAN" will be found a valuable collection of original papers and articles by nearly every leading writer on electrical matters, including :—

Prof. W. G. Adams,
Mr. G. L. Addenbrooke,
Sir James Anderson,
Mr Rollo Appleyard,
Prof. H. E. Armstrong,
Signor R. Arno,
Mr. L. B. Atkinson,
Prof. W. E. Ayrton,
Mrs. Ayrton,
Mr. J. Mark Barr,
Prof. W. F. Barrett,
Mr. W. W. Beaumont,
Mr. A. R. Bennett,
Mr. Shelford Bidwell,
Mr. T. H. Blakesley,
Herr O. T. Blathy,
M. A. Blondel,
Mr. Bertram Blount,
Prof. J. C. Bose,
Mr. C. Vernon Boys,
Mr. G. J. Burch,
Mr. A. Campbell,
Major P. Cardew,
Prof. H. S. Carhart,
Mr. E. T. Carter,
Mr. G. M. Clark,
Mr. W. R. Cooper,
Mr. R. E. Crompton,
Sir W. Crookes,
Mr. A. Dearlove,
Mr. P. B. Delany,
Prof. J. Dewar,
Herr M. Doliwo Dobrowolsky,
Herr Alfred Dubsky,
Dr. Louis Duncan,
Mr. J. Elster,
Mr. W. B. Esson,
Mr. Sydney Evershed.

Prof. J. A. Ewing,
Prof. C. A. Faure,
Prof. G. Ferraris,
Mr. W. C. Fisher,
Prof. G. F. FitzGerald,
Dr. J. A. Fleming,
Prof. George Forbes,
Prof. J. Frith,
Prof. W. Garnett.
Mr. J. Gavey,
Mr. W. Geipel,
Prof. H. Geitel,
Mr. A. H. Gibbings,
Signor G. Giorgi,
Dr. J. H. Gladstone,
Mr. R. T. Glazebrook,
Dr. George Gore,
Mr. A. Hay,
Mr. Oliver Heaviside,
Herr J. S. Hecht,
Prof. von Helmholtz,
Dr. H. Hertz,
Herr A. Heyland,
Prof. W. M. Hicks,
Mr. Paget Higgs,
Dr. Edward Hopkinson,
Dr. John Hopkinson,
Prof. E. J. Houston,
Prof. D. E. Hughes,
Mr. C. C. Hawkins,
Mr. Hugh Erat Harrison,
Prof. Dugald C. Jackson,
Dr. W. Jaeger,
Dr G. Jaumann,
Mr. Gisbert Kapp,
Prof. A. B. W. Kennedy,
Dr. A. E. Kennelly,
Mr. J. B. C. Kershaw,
Mr. Hamilton Kilgour,

Prof. P. Lenard,
Dr. J. Larmor,
Dr. S. Lindeck,
Dr. Oliver Lodge
Mr. L. B. Marks,
Prof. J. Clerk Maxwell,
Mr. P. V. McMahon,
Prof. Mengarini,
Prof. G. M. Minchin,
M. Henri Moissan,
Mr. W. M. Mordey,
Dr. Alex. Muirhead,
Prof. F. E. Nipher,
Mr. H. F. Parshall,
Prof. John Perry,
Mr. Nelson W. Perry,
Mr. C. E. S. Phillips,
Mr. W. H. Preece,
Dr. C. S. du Riche Preller,
Mr. W. H. Pretty,
Mr. W. A. Price,
Prof. J. H. Poynting,
Dr. M. I. Pupin,
Mr G. S. Ram,
Prof. W. Ramsay,
Mr. F. C. Raphael,
Mr. H. W. Ravenshaw,
Mr. J. S. Raworth,
Lord Rayleigh,
Mr. J. Rennie,
Mr. W. G. Rhodes,
Prof. A. Righi,
Mr. G. H. Robertson,
Prof. W. C. Röntgen,
Prof. Rowland,
Prof. Rücker,
Mr. Warren de la Rue,
Mr. Alex. Russell

Prof. H. J. Ryan,
Sir David Salomons,
Mr. W. B. Sayers,
Dr. Paul Schoop,
Mr. Louis Schwendler,
Mr. G. F. C. Searle,
Mr. J. S. Sellon,
Mr. Alex. Siemens,
Dr. Werner von Siemens
Mr. M. Holroyd Smith,
Mr. Willoughby Smith,
Mr. Albion T. Snell,
Mr. W. H. Snell,
Dr. W. Spottiswoode,
Mr. J. T. Sprague,
Prof. Balfour Stewart,
Mr. A. Still,
Prof. W. Stroud,
Dr. W. E. Sumpner,
Mr. James Swinburne,
Mr A. A. C. Swinton
Mr. F. A. Taylor,
Mr. Nikola Tesla,
Prof. Silvanus Thompson,
Prof. Elihu Thomson,
Prof. J. J. Thomson
Sir Wm. Thomson (Lord Kelvin),
Mr. H. Tomlinson,
Mr. A. P. Trotter,
Prof. John Tyndall,
Mr. E. J. Wade,
Mr. F. C. Webb,
Mr. F. M. Weymouth,
Mr. H. D. Wilkinson,
Mr. E. Wilson,
Mr. J. Elton Young,
Dr. Zetzsche,
&c., &c., &c.

And all papers read before the principal Electrical Institutions throughout the world by men eminent in the Electrical Profession have been given, together with authenticated reports of the discussions thereupon.

In addition to the above, "THE ELECTRICIAN" forms a complete record of all the important legal investigations which have occupied the attention of the Courts of Justice for the past 14 years, and it is customary for "THE ELECTRICIAN" to occupy a prominent position in the Courts as an authority upon all questions affecting the Electrical Profession. In this connection it is only necessary to point to the actions which have arisen from time to time upon the Edison and Swan Patents, the Compound Winding Patents, the High and Low Tension Systems, the Accumulator Patents, the Telephone Patents, etc., etc., in which "THE ELECTRICIAN" has figured as a reliable authority, and has been put in evidence and accepted by the parties concerned.

A regular feature in "THE ELECTRICIAN" has always been the verbatim reports of meetings of Electrical Companies and Corporations, and while "THE ELECTRICIAN" has never trenched upon the grounds legitimately occupied by financial journals, the information of a financial character given from week to week in its columns is full, reliable, and absolutely unbiassed. It has no interest whatever in any financial schemes, and is devoted entirely to the interests of the profession it was established to serve. "THE ELECTRICIAN" gives full reports of Meetings, &c., held on Thursdays, so that subscribers interested obtain their copies by Friday morning's post.

The original articles appearing in "THE ELECTRICIAN" are written by gentlemen having no interest whatever in particular electrical systems, and with but one object, and that the advancement of electrical knowledge and electro-technology generally. Many of these original series of articles have since been revised and amplified by their authors, and published in book form. These form the nucleus of the well-known "Electrician" Series, of which further particulars will be found herewith.

Finally, "THE ELECTRICIAN" has been of incalculable service to technical education, and has done much to make the general study of electricity the reality it has undoubtedly become. No aspirant to honour and renown in the electrical profession can hope to keep abreast of the never-ceasing stream of discoveries of new applications of electrical science to every day commercial pursuits who does not diligently peruse the columns of "THE ELECTRICIAN," which is pre-eminently the leading electrical journal.

TERMS OF SUBSCRIPTION:	YEAR.	HALF-YEAR.	QUARTER.
Great Britain	£1 6 0	£0 13 6	£0 7 0
Countries in the Postal Union	1 10 0	0 15 0	0 8 0

Bound Volumes (May and November), 17s. 6d., post free, 18s. 6d., abroad extra.

THE VERY BEST MEDIUM FOR ADVERTISING.

"THE ELECTRICIAN," SALISBURY COURT, LONDON, ENGLAND

www.ingramcontent.com/pod-product-compliance
Lightning Source LLC
Chambersburg PA
CBHW022058230426
43672CB00008B/1209